Cat Culture

The Social World of a Cat Shelter

In the series

Animals, Culture, and Society

edited by Clinton R. Sanders and Arnold Arluke

Cat Culture

The Social World of a Cat Shelter

Janet M. Alger
Steven F. Alger

Temple University Press
Philadelphia

Temple University Press, Philadelphia 19122

Published 2003
Printed in the United States of America

♾ The paper used in this publication meets the requirements of the American National Standard for Information Sciences–Permanence of Paper for Printed Library Materials, ANSI Z39.48-1984.

Library of Congress Cataloging-in-Publication Data

Alger, Janet M., 1937–
Cat culture : the social world of a cat shelter / Janet M. Alger and Steven F. Alger.
p. cm. – (Animals, culture, and society)
Includes bibliographical references and index (p.).
ISBN 1-56639-997-1 (cloth: alk. paper) – ISBN 1-56639-998-X (pbk.: alk. paper)
1. Cats–Behavior. 2. Cats–Social aspects. 3. Animal shelters. 4. Human–animal relationships. I. Alger, Steven F., 1941– . II. Title. III. Series.

SF446.5 .A36 2003
636.8–dc21 2002020417

This book is dedicated to the volunteers and resident cats at the Whiskers Shelter, past, present, and future

Contents

Preface

Why an Ethnography of a Shelter?

FROM THE BEGINNING of our marriage, our two cats moved with us through all the job changes associated with the early careers of academics. We are not sure how we advanced from our devotion to Georgy Girl and Alex to being animal activists, but in the 1970s we became conscious of the tremendous suffering of animals in our society. We began our activism modestly by contributing to local animal welfare organizations and feeding a stray cat in the alley behind our apartment. When we moved from Chicago to Albany, New York, in 1976, we did so with three cats, having added Casey to our family.

Soon after arriving in Albany, Alex died of cancer in spite of all our efforts to save him. We had no heart for new cats after this traumatic event, and it was not until we bought property in 1983 and found a mother cat and her four kittens in residence that we again became involved with stray animals. We found homes for two of the kittens and, as you might have guessed, added Margaret, the mother, and Nicholas and Annie, the kittens, to our brood. By this time the animal rights movement was well developed. We joined People for the Ethical Treatment of Animals (PETA) and followed events closely. We talked about the suffering of animals with our colleagues at work and fellow sympathizers, who we readily recognized. Our increasing concern for the condition of animals in general was bound up with the increasing number of cats in our household.

One day Janet heard about a stray cat who had given birth to her five kittens in the boiler room in a building on her campus. The sympathetic security staff brought her food and water and allowed her to remain. As the kittens grew, the cat family drew some attention as it moved around campus. The cats continued to live in the boiler room, entering and exiting through a broken window until the administration finally demanded that the broken window be repaired. The cats were left without a safe and warm place to stay as fall began to give way to winter. One of the security guards attempted the difficult task of rounding up the mother and kittens and tried to find homes for them. Because the cats had little human contact, gathering them proved difficult. He managed to capture four of the kittens, but not the mother and fifth kitten. He found a home almost immediately for one of the captured kittens, but he was less successful in placing the other three and took them in himself. Meanwhile, the kitchen and library staffs fed the mother and her remaining kitten. The mother supplemented their diet by hunting the large population of chipmunks on campus.

As the Christmas break approached, a campus librarian called Janet to ask for help in capturing and placing the two cats, because soon the campus would be empty. Having recently read a newspaper column about a local cat shelter that did not euthanize any of its charges, Janet contacted them. The shelter staff agreed to take the cats if they were leukemia negative. The intrepid librarian succeeded in capturing the cats though she suffered bites from the four-month-old kitten.

We whisked them off to our veterinarian, who used gloves to place them in a cage together, giving us our first close look at the pair. They were beautiful. They were both long-haired—the mother a brown tabby and the kitten a dilute calico. As a protective measure, the mother placed herself in front of the kitten, who peered at us fiercely over her mother's back. After sedating the cats, the veterinarian examined them and determined that they were leukemia-free. The next evening we took them to the Whiskers Shelter.

The shelter was located in a modest two-family house in a low-income area of town. The president (Regina) answered the doorbell and ushered us in to the dimly lit lower-floor apartment, which, to our amazement and delight, was full of loose cats resting intertwined together on old sofas and chairs. We had expected to see rows of caged cats forlornly waiting for adoption. Regina led us toward a very small separate room off the dining room, where, she explained, new cats were kept in cages. She placed our two cats together in a cage, where the kitten, once again, hid behind her mother. Regina then introduced us to Kate, whose father owned the house and who lived in the upper-floor apartment. She told us that Kate was gifted with cats and was the main human socializing agent for cats at the shelter. If anyone could make wild cats adoptable, Kate could, she said. Kate asked us to name the cats—a privilege of the one who brings them in—and told us that the names of related cats had to begin with the same letter. We named the mother Jennifer and the daughter Jesse. We looked on as Kate fed and began to try to relate to Jennifer and Jesse. Later, we sat among all the cats in the living room. We were very moved by the experience of various cats crawling into our laps for a petting. In spite of the crowding, it was obvious how much the people at Whiskers cared and how happy the cats seemed. So began our long association with the shelter in which we have been adopters (you guessed it, Jennifer and Jesse for two), cleaner/feeders, board members, and, eventually, participant observers in a sociological study.

We did not see a connection between animal issues and our work in sociology until some 10 years later when Janet began to prepare to teach a course on violence against animals in the college's Peace Studies Program. It dawned on both of us that there were many ways of bringing our interest in animals and sociological issues together. One of our first attempts at joining animals and sociology was in our study of multi-cat households, which focused on how caretakers understand the nature of their cats. Doing that study raised so many questions about the nature of cats that we were led to the present ethnographic study of the Whiskers Shelter. There we directly observed the interactions between the volunteers and the cats as well as the interactions among the cats themselves.

The peace studies course Janet taught was a form of activism as well as an intellectual experience. Learning to conceptualize animal issues sociologically helped us to see the connections between the animal rights movement on the one hand, and the civil rights and women's rights movements on the other. It helped us to draw parallels not only between these social movements but also between the oppression shared by minorities, women, and animals. At the end of the semester, Janet organized a student animal rights group on her campus, and we both joined a local animal rights group whose activities included protest marches against circuses and rodeos performing in the area. This book grew out of our increasing involvement on behalf of animals and in the work of the shelter.

The framework for our study is sociological: We used the theoretical perspective of symbolic interaction to interpret our findings and the ethnographic methods of both observation and participation to discover the underlying patterns of shelter life. Our intent was to expand our knowledge of animals from their own perspective. Such knowledge is necessary to devise humane policies for the treatment of particular animals. From a scientific perspective, we wanted to understand the nature of species of animals whose characteristics may vary widely under different circumstances. As two philosophers put it, we need to:

> Increase our understanding of the nature of non-human animals, especially their intellectual, emotional, and social lives. In Descartes' day, many people doubted that non-human animals had emotional, intellectual, or genuinely social lives. This view, however, has been abandoned by all but a few. Indeed, there is increasing evidence that animals have far more sophisticated lives than we would have thought possible. However, the evidence is less than definitive. We must conduct more systematic research. Although this will not solve the moral question, it will inform it.[1]

In this book, we describe the extraordinary social capacity of domestic cats as revealed in their everyday activities and relationships with the shelter volunteers and with one another. We describe the culture that emerged out of the interaction between the volunteers and the cats and show how the cats were active participants in the shaping of this culture. We also examine the culture that developed and was shared among the cats themselves–a culture that fostered equality, friendship, and social solidarity among the resident cats. And, we look at the special role played by the feral or "wild" cats at the shelter in creating and maintaining this distinctive cat culture. Finally, we present a picture of our shelter cats as having a sense of self and purpose in their social relationships with the volunteers and with one another.

Acknowledgments

WE THANK, first, the staff at Whiskers for their support throughout our long study. They tolerated our numerous questions, unsolicited advice, and our getting underfoot while they tried to do their work. We also thank the resident cats for their enthusiastic greetings, plentiful hugs and kisses, and unfailing interest in our activities.

Janet thanks her colleague Ned McGlynn for providing the opportunity to teach a course on violence against animals in the Peace Studies program as it was that course that helped us to link our animal activism with sociology. She also acknowledges the library staff at Siena College for their research assistance, particularly Catherine Crohan and Sean Maloney. She is grateful to the Teaching Committee at the college for supporting a summer grant to purchase qualitative analysis software. She also thanks colleagues Doug Fraser and his wife Kathy for their comments on some initial findings of the study and Bob Woll for offering his wisdom.

Steve thanks the College of St. Rose for release time to work on the manuscript and the Professional Development Committee at the college for a mini-grant to purchase qualitative analysis software. Thanks also to his graduate assistant, Puspita Sen, for her excellent research support and insights into the human–animal bond. He is particularly grateful to his colleague and friend, Richard Wunderlich, for his encouragement and support throughout this endeavor.

We both thank Wendell Bell of Yale University for, as always, encouraging us to continue with our research even though it was then more outside of the scope of sociological endeavor than it is today. Russ Schutt of University of Massachusetts, Boston, also helped up by suggesting some key literature from neurobiology. Carol Hall carefully read the first draft of the manuscript, offering many helpful comments and clarifying points of Whiskers's history for us. Clint Sanders of the University of Connecticut was our initial source of inspiration to proceed with this study and has been a major source of support to us throughout. Janet Francendese's editorial assistance was an invaluable part of writing the book. Our thanks go, as well, to Ellen Johnson, whose secretarial expertise shaped the manuscript for presentation.

Finally, this would not have been possible without the inspiration of our own wonderful cats through the years: Georgy Girl; Alex; Casey McGruder; Margaret and her children, Nicholas and Annie; Jennifer and her daughter, Jesse; Sydney; Calvin; Shelly; Amanda; Cassidy; Petey; and Amberlee.

Portions of this book are adapted from previously published articles: "Beyond Mead: Symbolic Interaction between Humans and Felines," *Society and Animals* 5(1997): 65–81, with permission of Brill Academic Publishers; and "Cat Culture, Human Culture: An Ethnographic Study of a Cat Shelter," *Society and Animals* 7(1999): 199–218, with permission of Brill Academic Publishers.

1

The Myth of the Solitary Cat

"CATS . . . LIVE in our homes without any attempt to conform to our standards; they pursue their own agenda, they cannot be relied upon to share our feelings, their minds are less open to us, and they seem quite immune to human or canine guilt."[1] Here, Katharine Rogers captures the most common understanding of the domestic cat in our culture. As sociologists, we would call her statement a social construction; that is, humans in our culture view cats as having these characteristics quite apart from any knowledge of their innate capacities and regardless of contrary evidence and personal experience. The power of such constructions lies not in their accuracy but in their ability to give meaning to our own collective lives in a particular period. Thus, in our culture, cats are mysterious creatures whose motivation comes from within. Aloof, independent, and solitary, they lack a strong interest in human affairs and cannot be counted on to share our feelings or give solace. Their own lives tend to be centered on instrumental activities such as hunting and eating. They are never fully tame or fully committed. The unspoken contrast here is always to our other popular animal companion, the socially constructed dog, who lives largely for us and willingly conforms to our standards.

In Western culture we have made a place for companion animals since the eighteenth century. In the seventeenth century, the philosopher Rene Descartes declared all animals to be soulless and unworthy of concern. In reaction, the French upper classes of the eighteenth century developed an alternative, humane perspective on animals from which cats benefited. The cat came to be included with dogs as a companion animal and the literate began writing about and expressing their feelings toward their animals. Enlightenment philosopher Jean Jacques Rousseau was fond of cats and noted that domineering men did not like cats "because the cat is free and will never consent to become a slave. He will do nothing to your order, as the other animals do."[2] This was clearly intended as a criticism of domineering men and a compliment to the domestic cat, who symbolized the new emphasis on liberty and independence. The French influenced the English in this new humane sensibility toward animals. The famous eighteenth-century man of letters, Horace Walpole, wrote frequently about his own animals and those of his friends. One of his cats, Selima, was immortalized in an ode written by his friend, the poet Thomas Gray.[3]

The eighteenth century is clearly a time of transition in the stature of animals. In writing about one's feelings for animals one had to be humorous, mocking, or at least apologetic to save one's reputation. By the nineteenth century, writers were no longer discouraged from discussing the cat and other companion animals as worthy friends with their own personalities and endearing traits. Writers downplayed their wildness and independence and emphasized their sweetness, fastidiousness, and domesticity.[4] Mark Twain, in *Pudd'nhead*

Wilson, describes the houses of Dawson's Landing as having window ledges often adorned with a cat "stretched at full length, asleep and blissful, with her furry belly to the sun and a paw curved over her nose. Then that house was complete, and its contentment and peace were made manifest to the world by this symbol, whose testimony is infallible."[5] Cats frequently were also included in paintings of domestic scenes in the United States, Great Britain, and France.

In the industrialized twentieth century, with its unending restraints on our freedom, writers again emphasized the wildness and independence of cats as positive and laudable characteristics. The cat in Leila Usher's poem, "I Am the Cat," written in the early 1900s, speaks to us:

> Could they but see themselves
> As I, the Cat, see them,
> These human creatures, bereft of all freedom,
> Who follow in the ruts others made
> Long ages gone!
> Who have rings in their noses,
> Yet know it not.
> They hate me, the Cat,
> Because, forsooth, I do not love them.[6]

Rogers, who has studied historical changes in feline imagery, notes that later twentieth-century writers have been successful in taking on the perspective of cats: "To do so requires not only sensitive interpretation of observed feline behavior . . . but an acceptance of the animal as a quasi-equal being whose feelings and claims can be presented as important without being falsely inflated into human ones. Twentieth-century writers are more ready to recognize that cats and other animals are individuals with lives of their own, and thus can create a cat's consciousness, seemingly free of projected human values."[7] Equality, then, is a significant addition to the contemporary social construction of the cat. To what extent can we say this greater openness to interspecies communication has led to a more "objective" view of the cat?

Rogers sees this intersubjectivity as affirming and accepting the traditional social construction of cats as aloof and solitary. Cleveland Amory's popular books about his famous cat Polar Bear are typical of current attitudes, she argues, "in celebrating their subject's amusing perversity rather than his sweet affection or refined ways. We now admire in cats the very qualities for which they were censured in the past–their refusal to comply with human wishes and standards, as well as their emotional independence and their cool pursuit of self-interest."[8] On the other hand, Rogers gives many other examples that indicate quite different feline characteristics, which she either ignores or discredits as too dog-like. For instance, she accuses an author of a cat autobiography of making the cat into a dog by giving it a name so the cat can have self-respect and be able to come when called![9] In another instance, she speaks of altruism as unfeline.[10] In short, Rogers discounts as sentimentalizing any instance in which cats are described as more affectionate, committed, or obedient than the accepted dominant social construction. Accounts like those of Rogers cause us to wonder whether there are alternative perspectives on cats. The cat portrayal in more popular works or in the minds of cat owners is an altogether different cat.

The Unsung Cat

Let us begin with the subject of cat heroes. The very idea of cat heroes seems remarkable because it contradicts the common perception of cats. Swanson, the author of a book on cat heroes, notes, "The qualities most commonly associated with the feline species include selfishness, indolence, cunning, and a tendency toward pique. According to popular opinion, cats extend themselves only insofar as they see an opportunity for a meal or fear the loss thereof."[11] Nevertheless, there are many documented cases of cats who risked their lives or who took extraordinary steps to save a member of their own or other species. Among the many instances of mother cats who have risked their lives to save their kittens is the story of Scarlett in New York City. Scarlett returned five times to an abandoned auto shop engulfed in flames to save each of her kittens. A firefighter found her badly burned and brought her and her kittens to the North Shore Animal League, where all but one of the kittens survived. The mother's recovery was lengthy, and she will require lifelong medical attention. Sociobiologists would regard Scarlett's behavior as neither loving nor altruistic; it is merely the action of the selfish gene clamoring to live on in the next generation. The selfish gene evidently is not very intelligent; Scarlett was too badly burned to have cared for her four-week-old kittens who were too young to survive without her. It would have made much more sense genetically for her to have saved, say, two kittens and written the others off so she could have continued to care for the ones she saved. But firefighter Giannelli noted, "Though her eyes were swollen shut and her paws burned the cat made a head count of her young ones, touching each kitten with her nose to make sure they were all there."[12]

The numerous instances in which cats have saved children are even more difficult to explain in genetic terms. Mr. Meow was five years old when the Peckroads brought their first child home from the hospital. They had prepared the cat for the big event in a variety of ways, including allowing him to visit the baby's room as they decorated and stay close by as they changed or rocked the baby when she came home. Several months after the baby's arrival, Mrs. Peckroad heard Mr. Meow yowling frantically. She began to search for him and realized that the cat's cries were coming from the baby's room. She ran into the room to find Mr. Meow on a dresser looking into the crib. "He turned and uttered a piercing cry that brought her rushing to the side of the crib–where she saw, to her horror, that the mobile [she had recently placed over the crib] had fallen into the crib, and baby Sam had gotten tangled up in the cords."[13] Sam was already blue and had to be rushed to the hospital for treatment. Had it not been for Mr. Meow, the baby might have died.

Cats have also saved dogs. About a year after Surya and Matt Drummond took in Tramp, a stray cat, they brought home a golden retriever puppy they called Lady. When Lady was barely grown, she slipped out of her collar and disappeared. The Drummonds searched the woods and fields around their home until dusk with no success. When they returned home, to their dismay they could not find Tramp either though they normally kept him in at night. To try to attract both animals, they put food on the porch and, as they scanned the area, they noticed two glowing eyes at the edge of their property. Matt went to investigate and saw that the eyes belonged to Tramp, but no amount of coaxing would get him to come home. Surya, then, brought the food to him to entice him, but, upon seeing her, Tramp headed into the woods. The Drummonds followed him and finally heard the whimpers of Lady, caught in a fox trap. Tramp took a strong interest in the rescue, even jumping into the car and accompanying them to the vet to treat Lady's broken leg.[14] Anyone who has tried to take a cat to the vet will know that this is, indeed, extraordinary.

These examples of feline heroism should at least raise questions about the social construction of cats as uncommitted and lacking in empathy for their significant others. The story about the dog rescue is particularly interesting because the owners indicated that Tramp and Lady were not close. They pretty much went their own way under normal circumstances. But, in an emergency, Tramp acted toward Lady as if he recognized Lady as a member of his group to whom *he had a duty*.

Our own experience with cats has also been quite at odds with common cultural perceptions. We have lived with 15 cats over the last 34 years and have kept notes on their behavior during various periods. Only one of our cats (Calvin) clearly demonstrated unaltruistic behavior. When he first came to live with us, he was disinterested in us, had his own agenda, and lacked empathy for us and our other cats, but he has moved away from this behavior over time. The other 14 have differed from this social construction in key respects. In particular, our cats have been quite sociable and involved in human activities. Although the houses we have lived in have been spacious, the cats have all wanted to sit in whatever room we occupied, often on our laps. Frequently they try to sit on the work in which we are engaged. Several have insisted on sleeping with us, and all have wanted to investigate whatever we do. Some were vocal from the beginning, but all of them became more so over time. They have greeted us enthusiastically in the morning when we have arisen and when we returned home, particularly after we have been away. There has not been a true loner in the lot of them.

They have all wanted to include us in their activities, making demands on us to play, pet them, or see a mouse they have caught before they would eat it. Frequently they would leave their catch for us to admire. When we lived in an apartment in Chicago and had only two cats, Georgy and Alex, we took them down to the yard in good weather. We would let them out the back door and follow them down the stairs. If we returned to the apartment to retrieve something we forgot, we would find them waiting for us on the landing or the stairs. They clearly defined the trip as a joint experience.

We have also found them to be highly adept at manipulating their environment, including us, to better serve their needs. We have been startled to find how many more interactions they initiate than we do. Whether one looks at instrumental interactions such as the quest for food or expressive ones such as the quest for affection, we would have come up with a very different social construction of the cat. Our cats have been highly sociable, involved in our activities, and quite dependent on us.

They are also very sociable with one another.[15] They greet one another by touching noses, sit together, imitate one another, and play together. Some groom each other. Shelly picked up Calvin's habit of drinking from the faucet, which she had not done until she saw him doing it. There are numerous examples of playing together, but we will give just one from our notes:

> Cassidy is full of beans, biting and kicking the area rug in the family room. Calvin comes in and she attacks him. He squeals and gets in defensive position, looking very serious. She circles him and he keeps turning to face her and she attacks three times, each time bringing squeals or grunts from Cal. Then

> she flops and rolls about as he watches. Something in another room catches her attention and she turns her head. She suddenly thrusts her front legs, claws out, in his direction. He gets up and puts his front paw out in her direction. He does this when he wants her to attack him. She gets up and half-heartedly obliges him. Their bodies meet in a stand-up-on-hind-legs position. They move down to all fours; she jumps over him and climbs to the upper tree seat on their cat tree. He watches her. She shifts to the radiator. He sits near the stereo cabinet and they look at each other. She goes to the window. I let her out into the cat run. He follows.

Calvin is very fond of Cassidy, who came from the Whiskers shelter. He allows her to sit in his favorite spots and, unlike his treatment of other cats, he does not slap her and try to make her leave a place he wants to occupy. He loves it when she attacks him and often tries to entice her to do so. She does not always oblige him. Their facial expressions are very communicative; when Calvin wants Cassidy to attack him and she is contemplating doing so, they both tuck their chins in toward their throats and stare intently at each other. Then, Cassidy will look away for a second while Calvin continues to stare. This is generally the signal that the attack will commence.

In short, if our information were based solely on our own cats, we would have developed a very different construction of the cat than that which exists in our culture. This alternative portrait was given further credence in our discussions with other "cat people" we met through the Whiskers Cat Shelter and in our observations of the shelter cats. As a result of our observations, we decided to study multi-cat caretakers and their cats.[16] Because human–cat interaction varies greatly with the nature of the household, we deliberately chose 20 multi-cat owners who were either single or married without children living at home. These owners were committed to their cats and made time for them. Under these optimum conditions, we could ensure that the cats would act freely. These would be homes in which cats received a great deal of positive attention from their human associate(s) and had relationships with other cats. In addition, we felt that studying multi-cat households would allow us to compare the distinct personalities of the several cats. If, in such households, owners perceived their cats as aloof, uninterested in human affairs, lacking in empathy, and solitary, we might truly say that there is strong evidence to support the prevailing social construction of the cat.

Through the Eyes of Cat Lovers

We apply symbolic interaction to our study of cat interaction with humans and with other cats. The sociological perspective of symbolic interaction treats humans as active constructors of their social world: Humans do not act solely on the basis of norms and other external constraints. Rather, human actors receive and evaluate a social stimulus in terms of their own goals and prior experience as well as in terms of the social norms. The actor's *subjective viewpoint* thus is a factor that must figure in any explanation of how, once a stimulus occurs, actors define the situation, select a course of action, and act. For actors to interact symbolically, before they make a choice they must be able to imagine how others perceive them and how others might react to their choices. They can imagine the meanings others will attach to alternative courses of action because they can take the perspective of the other. This *intersubjectivity* becomes the basis for cooperation and community as well as opposition and rebellion. Humans always have a choice. Consider the student who is turning a paper in late because he went to a party instead of working at the library. Which approach should he take with this professor? Should he tell the professor that his dog ate his paper or his printer ran out of ink? Or should he just tell the truth and beg for mercy? He knows this professor does not like lame excuses. He decides to try for mercy.

George Herbert Mead, a central figure in the development of symbolic interactionism, asserted repeatedly in his work that non-human animals could not engage in symbolic interaction. He believed that they lack the required cognitive skills, such as memory of past events and ability to project into the future, and they cannot take the perspective of the other and imagine how the other would respond to one's own actions. Most significant to Mead is animals' lack of language and, thus, the inability to converse with oneself about alternative courses of action without acting them out. Without language it is not possible to have a sense of self and see oneself as others do. According to Mead, animals may communicate with one another through gestures, but there is no indication that they are aware that their behavior has meaning for other animals. Further, they have no control over their gestures, which are instinctual manifestations. Animals, then, are not engaging in symbolic interaction when they communicate through gestures, because something is symbolic only if it is under one's control.[17] Thus, the play we described above between Calvin and Cassidy would not be meaningful. It would just be the acting out of some instinctual imperative.

Mead's famous colleague, Charles Horton Cooley, did not see language as critical for symbolic interaction. He wrote about his daughter's ability to take the perspective of her mother by the time she was six months old, long before she developed language skills.[18] Recently psychologists have begun exploring these issues and argue that even infants two or three months old have some capacity to take the role of the other in a social relationship, forming the basis of a complex mutual understanding.[19] We would argue that such role-taking is based only minimally on cognitive development. More important are the infant's observational skills, which provide the information about the other, and the social bond between infant and caretaker, which provides the motivation to understand and respond to the significant other.

Contemporary sociologist Randall Collins also argues for an emotional dimension to the concept of role-taking. He suggests that there are two types of symbolic interaction differentiated by type of goal: *practical goals* generated by our relationship to nature, such as problems of survival, comfort, and so forth;[20] and *social goals* generated by our relationship to social groups and focused on symbols of solidarity. Social goals develop out of what Collins calls "natural interaction rituals," which require at least two participants in the same location who "focus attention on the same object or action, and are aware that each other is maintaining this focus," and they "share a common mood or emotion." This creates a shared reality in which "participants feel like members of a little group, with moral obligations to one another. Their relationship becomes symbolized by whatever they focused upon during their ritual interaction. Subsequently, when they use these symbols [such as a secret handshake in a fraternal organization] they have a sense of group membership."[21] Symbols remind them to reconstitute the group assembly. Such social goals are expressive in nature and valued for themselves rather than having any utility.

The inner lives of animals and their relationship with one another become viable topics within this revised concept of symbolic interaction. We may ask whether something so evolutionarily advantageous would be likely to develop only in humans and not in other related animals who also face situations in which competing courses of action appear.[22] We must revisit the question of whether animals have a sense of self, a personality. We must ask whether animals choose between competing courses of action and how they make such choices. Through what mechanisms do they take the role of the other? Are emotional attachment, smell, and/or relational thinking involved? What types of symbolic interaction do animals engage in? Are shared meanings and a sense of past and future possible? Sociologists are now beginning to raise these questions.[23]

Following Sanders in his study of dogs and their caretakers, we asked our feline caretakers questions based on four issues. First, we tried to establish whether they saw their cats as able to think, including the ability to anticipate actions, solve problems, and make choices between alternative courses of action. We wanted to know if the caretakers viewed their cats as individuals with distinct personalities. We asked the owners if they saw their cats as emotional, empathetic, and able to reciprocate in the relationship. Finally, we wanted to know if they considered their cats to be family members with social standing in the home and the right to be included in various social rituals.[24] As caretakers of cats ourselves, we were not surprised that their responses did not conform to either Mead's views on animals or the social myths surrounding cats.

The Cat as Minded Actor

Virtually all of our respondents answered in the affirmative when we asked them the question, Do you think your cats think? We then went on to explore different dimensions of thinking. To see whether the caretakers believed that their cats could assess the future, we asked respondents if their cats anticipate events. One owner's husband told her that her favorite cats go to the door and wait for her shortly before her customary time of return from work. Another said that one of her cats knows when he is going to be medicated and hides. In our home, Shelly knows when we are going to brush her long, beautiful fur and runs away and hides before we can catch her. It also appears to us that some of our cats know when they will be taken to the veterinarian and hide *before* we take out the carriers we use to transport them. Since we do not take the cats to the veterinarian at frequent or regular intervals, it suggests that the cats can read our demeanor. These anticipatory actions are important evidence of the cat's ability to take the role of the other in social interaction.

We also asked the owners if their cats ever "figured out" something the owners did not expect them to grasp. We received numerous examples of such behavior. One respondent believed her cat learned on his own to ring the doorbell when he wanted to come back in the house. Another thought her cat had figured out how to flush the toilet to scare another cat away from her food so he could eat it. This owner also believed this same cat saved her life by sitting on her chest to wake her up when the furnace gave off toxic fumes. Almost all of our owners told of their cats' ability to open doors and cabinets. Thus, they believed that problem-solving was well within their cats' mental capacity.

We then asked our respondents if their cats ever seemed to make choices between alternatives. Such situations are treated as important in the symbolic interactionist conception of thinking because they require an internal conversation about future consequences. All of our owners were able to recount such situations. Several told us that their cats would wait to eat the food placed in front of them until they were sure they could not get the owner to give them something they liked better. As one owner described it, "Honeybun, when I first give her the food she is supposed to eat (Hills), will wait a bit to see if I will give her favorite (Fancy Feast); eventually, she will eat the Hills food."[25] Another study participant talked about her cat weighing alternatives:

> Piccolina (the blind one) also loves to go outside, but I can only let her do it when she's going to have my undivided attention, because she'll march right down the street without hesitation if I don't stop her. So when I do let her out, I sit on the back steps watching her. She'll spend time in the grass, etc., within her allowed range, but when she hasn't heard my voice in a while, she heads toward the driveway. The farther she gets, the faster she goes, as if she knows her chances of escape are improving by the inch. Then I say her name in a very low, demanding voice; she stops. Then I tell her in the same voice. . . . "Get back in the yard." She knows just what I am saying, and she turns around very grudgingly and walks back toward the yard, making little noises of protest along the way. I think of it as grumbling. She definitely knows she's not supposed to go down the driveway, tries to do it anyway, and knows when I tell her to go back.[26]

Although the caretakers did not believe that the cats had conversations with themselves in human language, they gave examples of cats appearing to make mental calculations based on memory, taking the role of the other, and assessing future consequences. These mental calculations allowed the cats to define the situation, choose a course of action, and change that course when necessary.

The Cat as an Individual

All of our cat caretakers saw their cats as individuals with distinct personalities. They were able to describe each of their cats in terms of such characteristics as his or her temperament, demeanor, playfulness, talkativeness, intelligence, and forms of self-expression:

> Mai-Ling is boss of the house–from Whiskers–independent and assertive. She established dominance over James Joyce [another cat] early on. She has no sense of humor. Katie [another cat] annoys her.
>
> Jasmine is a tomboy. If she were human, she would wear blue-jeans and get dirty. She's defiant and strong-willed from a kitten. She has a very colorful personality. She'll scream at you if she doesn't like what you're doing. . . . She's more in tune with John's [husband] emotions. John gives her free reign of everything. Jasmine rides on John's shoulder around the house.
>
> Chaucer is an extremely aggressive male. Like the school bully. Has to be top cat. Must have own way or he'll bite. . . . Headstrong. Can't physically stop him. Doesn't bite as hard as he used to–just enough to let you know who's boss. . . . Chaucer calculates–watches–very observant.[27]

The caretakers were particularly likely to emphasize how affectionate their cats were, how they related to the owners themselves and to the other cats in the household, and how they reacted to visitors. One respondent described one of her six cats in these terms: "Honeybun is the biggest love-mush. She loves everyone. She does not relate to the other cats but she cuddles with everyone who comes into the house. She demands affection and will actually 'hit' people with her paw to get them to pet her or to keep petting her."[28]

Many of our owners believed that their cats had strong likes and dislikes that added to their individuality. These preferences centered on a variety of things such as food, play, and social relationships: "Alfonse is crotchety.... He likes to knead hair. He likes peas and lima beans and corn." "Chaucer loves food and a dripping faucet." "Sheepie dislikes being told what to do. Gets angry if she wants to go out and we say no. Adores being with us outside. Loves the outside in general." "BC likes string; you can always lure her out of a hiding place with string. She also likes to sleep under the stove when there is a fire in it."[29]

Our cat caretakers often distinguished their cats in terms of their histories. Many of their cats had been rescued, either by the owners or by shelters, from situations of neglect or abuse. This background information was used to explain such traits as fear of strangers or an unaffectionate personality. A cat's history also entered into the caretaker's description of how the cat changed over time. One respondent described her cat who had been rescued from a collector as "shy, but starting to become playful and trusting."[30] Another said that her Kate:

> Was a stray, abused. She was a long time coming to me. Can only pet her from the back; if you try to

> touch her from the front she shies away. She approached my husband sooner than she did me. She plays by herself and with Tharpin [another cat]. She sleeps on the bed. . . . She doesn't really know how to play; she watches the others play. There was no one to teach her how to play.[31]

We conclude that these multi-cat caretakers saw their cats as individuals whose characteristics were affected by their history and experience as well as their inherited characteristics.

The Cat as Emotional and Reciprocating

Although we did not focus centrally on cats' emotions in this study, all of the respondents commented about their cats' feelings. They believed their cats showed such emotions as joy, love, sadness, jealousy, and fear and anger. Since cats in homes in which their basic needs are satisfied may experience a larger range of emotions than domestic cats living on their own, we will return to this topic later. In the realm of feeling we focused on whether cats were responsive to the emotions of their caretakers. Almost all of our respondents perceived their cats to be empathetic, as indicated by the following typical comments: "They are there for you. If you feel bad, they are always there. They can read your feeling. Bruce–if I am feeling bad–gets real close to me–very sensitive." "They have a sense of when things are not right. When we are sick or upset, they come to us and stay around us." "They sense my moods. If I am sad or crying, Cabbage comes and rubs against me." "James knows my moods. He can tell if I am upset and will climb in my lap. If I am crying he will lick my tears."[32]

Our caretakers also thought their cats were intensely interested in their affairs, following them around the house, investigating what they were working on, sitting on their work, greeting guests, and involving themselves in every aspect of their humans' lives. Further, as has been our experience, the owners thought their cats initiated many interactions. The cats were active in molding the relationship to and for their satisfaction: "Claudia comes up purring and howling to be picked up." "When Sheepie gets bored she tries to initiate play by being underfoot. She starts batting something around the floor all around us." "Patches cries to go outside. If we are outside he will cry at the screen . . . till we let him out. It always works. Shayda will hit you to pet her." Some cats affected their owner's responses to other people: "Once a pizza delivery man came, and the girls [cats] growled and stalked him and we got rid of him fast. They knew something was not right."[33]

We were alerted to an important aspect of human–cat interaction because of caretaker comments about the empathy of their cats and the cats' high involvement in their lives. The caretakers were motivated to interact with their cats by social expressive goals. That is, the relationship with the cats was an end in itself in which the owners derived pleasure from the interaction for its own sake. The interaction of the cats toward their humans was both instrumental (a means to an end) as well as social and expressive. For the animals, the human controls both their instrumental and expressive needs. We expect that the more stable and regular the meeting of instrumental needs such as food, water, and a safe environment, and the more attached the cats are to the owner, the more emphasis they will place on expressive needs such as play and affection in interaction with the owner. The cats in this study were well loved and cared for and allowed out only to a limited extent and under supervision. Their interactions with their caretakers were dominated by their desire for affection, attention, and play.

In his study of dogs, Sanders focused on the emotion of guilt.[34] His study participants "saw their dogs as possessing a basic sense of the rules imposed by the human members of the household."[35] When the dogs violated these rules, their owners perceived them as showing guilt. This reciprocating behavior of dogs in relation to rules set by owners probably results from the fact that dogs require considerable training to be acceptable household members. The dog owner and dog, then, must focus significantly on practical goals. We specifically asked our respondents what rules they set for their cats and whether they attempted to train them. They provided comical responses. First, they set few rules and these were for things in their complete control. For instance, most said the cats could not go out, but this was not a rule in the sense that they opened the door and the cats could choose to obey it or not. Other rules related to scratching the furniture and climbing on counters and tables. These rules were not important to the owners. Some did not try to gain compliance, whereas others made half-hearted attempts. One claimed she would be tougher, but her husband was overly indulgent and allowed the cats to do what they wanted: "They are not allowed on the table by me but David lets them. They don't listen." "It used to be that they weren't allowed on counters, but now because of Rosie [a cat] living in the kitchen it can't be helped. They can't eat from my plate while I'm there. I used to chase them from the furniture, but now the furniture is unscratchable." One owner would have preferred them not to scratch the furniture but said, "Scratching training is hopeless. We never tried."[36]

Since owners did not make rules a major aspect of their relationship to their cats, they did not invest much effort in enforcement or training. A couple of caretakers made half-hearted attempts at training, and others thought it was not possible to train cats. Many of them actually opposed the training of their cats. As one woman said, "Training? No, I don't see it as a cat thing. I know you can, but one of the great things about cats is they are who they are. It's not about training. If I were going to train something, I would get a dog." Another explained, "No. Never. Cats are free spirits. I want them to be as much of a cat as they can be."[37] As these owners see it, training is a violation of the nature of the animal and, even though it is possible to train a cat, it is undesirable. Moreover, many see their cats as training them rather than the reverse: "They have trained us to give them food when *they* want it, pet them when *they* want to, and play with them."[38]

Rather than focusing on practical goals, caretakers interacted with their cats for social reasons. Cat owners seemed to be the most satisfied with their relationship to a particular cat when the cat initiated more or an equal number of their interactions–that is, when the cat reciprocated love and friendship. We cannot find evidence of the cat's ability to take the role or perspective of the other in training situations. Rather, we must find such evidence in the owners' descriptions of their cats' empathetic responses toward them and in the reciprocity of play behavior. One of our caretakers described her cats as little thieves who steal things such as socks and then bring them back to her or her husband when they want to involve them in a game. Clearly, these owners see their cats as "knowing" both they and the cats are playing.

Affording the Cat a Social Place

All but one of our cat owners saw their cats as authentic family members. Many of our respondents were either single or married women without children who tended to see their cats as children. One explained, "They are my children. I can't imagine my life without them. I would never give them up for anything. I have made provisions for them in my will if I should die."[39] Another respondent who had recently married said, "They are an integral part of the family. I consider their feelings. When I moved I was very concerned about their adapting. I would never have married a man who didn't want them or moved where they couldn't go."[40]

The cats were also considered a part of family routines and many built their daily activities around their cats. Typical examples were getting up in the morning, feeding and grooming, playtimes, supervising the cats outdoors, and going to bed. One respondent described the latter routine with her six cats: "At 8 P.M. they all get a treat of shaved turkey breast. At bedtime they all get "Cluckers" (a cat treat). They come in the bathroom while I brush my teeth and then we all troop down the hall to the bedroom where we all sleep together."[41] The cats know these routines and remind the owners if something intervenes. One owner said, "They know when it's bedtime. They gather of their own accord. If I'm not ready the dog and Alfonse [a cat] come and get me."[42]

Family routines are examples of "natural interaction rituals" in which participants are mutually aware of focusing attention on the same object or action.[43] These objects and/or actions then become the basis for the emergence of collective representations that, in turn, can be called upon in the future to regenerate the common mood and bond that unite participants in a shared reality. The woman in the previous quote created the shared reality that it was bedtime for her and her cats by distributing "Cluckers" and making the trip to the bathroom. This symbolic interaction was directed at social goals.

The Social Cat and the Shelter Study

Our study of multi-cat owners gives little support to the myth of the cat as aloof, independent, and solitary. The cats in our study were highly social. They did not pursue their own agenda but were part of a group of other cats and humans in which routines and standards of conduct were mutually derived at through an interaction process. Their owners perceived them as empathetic and highly responsive. Only in the realm of rules and training did the dominant myth hold for them. But anyone who has attended a cat show, a cat circus, or a performance at the pier in Key West of domestic cats jumping through hoops of fire, for example, knows that cats are eminently trainable. We have personally had great success training our cats not to scratch on furniture. It took very little effort or stress for us or our cats. Of all the cats we have cared for, we found only one cat to be challenging to train because he took longer than normal to comply. We think that the opinions about training among highly committed caretakers say less about cats than they do about our needs as humans in a highly constraining society. On the other hand, those who have observed cats to be aloof, independent, and uninterested in human affairs are not necessarily wrong. Cats are highly adaptive creatures who adjust their behavior to different situations. People who believe that cats are aloof, treat them as if they were and reinforce that potential within the cat. Our multi-cat owners saw cats in quite a different light, treating them as if they were loving, empathetic, and important in their lives and thus reinforcing a different set of cat capacities.

These owners saw their cats as highly individual beings who differed in temperament, demeanor, playfulness, talkativeness, intelligence, inventiveness, expressiveness, and displays of affection. They had strong individual preferences in food, sleeping arrangements, and games. They were more drawn to some of their fellow cats over others and to particular humans as well. They spent a great deal of energy making their desires known to their human companions in a wide variety of ways, giving them no peace until they were understood and their needs satisfied.

Our owners saw their cats as taking the role of the other, defining situations and selecting courses of action. Cats and caretakers experienced shared meanings, which are at the heart of symbolic interaction. We suggest that such shared meanings were derived from the cats' close observation combined with their affection for their caretakers and are not necessarily linked to common human-type language. Many of our respondents commented on how closely their cats observed them. We are also aware that our cats watch our facial expressions and our actions, and show listening behavior by cocking their ears before we speak. The combination of the close observation and emotional attachment allows the animals to understand the caretaker's perspective and choose between competing courses of action. It also allows shared meanings to emerge as all parties attempt to convey a message to the others. Finally, it allows for the development of collective representations or symbols that unite cats and caretakers in a shared reality. All of this can be accomplished without the use of language.

Nonverbal communication, then, is central in understanding the relationship between humans and their nonhuman companions. Indeed, most communication is nonverbal. We tend to be unaware of this until we try to communicate with a child, someone who speaks a different language, or a person with a disability that limits speech. Sociologists are most likely to observe this type of communication in ethnographic research rather than in surveys that focus largely or entirely on words. According to Arluke and Sanders, "The accuracy of our understandings of the other–human or nonhuman–is grounded primarily on our *sense* of his or her body and behavior. The validity of this understanding is confirmed or denied by its practical outcomes. The 'truth' of our perceptions and expressions of the other's orientation derives from whether these understandings work to establish communication and sustain a viable and mutually rewarding flow of interaction."[44]

Finally, based on our own evidence and that found in other studies, we strongly suggest that far from being a human attribute, symbolic interaction is a widely distributed ability throughout the animal kingdom that enables animals to survive more effectively in varied environments. Even male guppies will watch other males to see which ones the females prefer and which ones they avoid. Then, if given a choice, the observing male will swim near the rejected males, presumably so that the females will be more attracted to him instead.[45] As more and more scientists reach this conclusion and recognize that these factors provide more valid information than behaviorist and instinctual presumptions, perhaps the gap between scientists and those who have spent a great deal of time with animals in relatively natural settings will decline. If animals are capable of symbolic interaction, then the many forms of human–animal interaction that exist in societies and interaction among animals themselves are appropriate subjects for sociological research.

We have transferred the key questions in our first study to the Whiskers Cat Shelter, and have extended them to look at the social life of the domestic cat as it has developed in the context of the shelter. We reasoned that if cats can engage in symbolic interaction they might well have a culture composed of norms, sanctions, and, perhaps, even roles shared by the group. To discover whether such a culture exists, we used our work as volunteers at the Whiskers Cat Shelter as the foundation for a research project. The chapters that follow explore the social world of the shelter.

2

The World of Whiskers

WHISKERS DID NOT originate as a shelter. It began in 1982 when Jane Donne opened a storefront thrift shop to earn money to spay and neuter stray cats. It was not long, however, before people began leaving cats at her door. Jane took them in, of course, and tried to find homes for them. During this period, Megan, a fellow cat lover who lived in the neighborhood, visited the thrift shop and became very interested in Jane's efforts to rescue stray cats. She began helping Jane in various ways and would soon play a central role in the evolution of this effort that, as yet, had no name.[1] As more cats arrived, Jane moved to a larger store with more space in the back to house them. There she met Kate, an upstairs neighbor, who was very sympathetic to Jane's efforts. Other neighbors, however, complained about the cats, and in 1986 Kate helped her rent an apartment that became the first official shelter site. One evening in Albany during the preshelter period, Jane was trying to trap a stray cat when Regina passed by and thought she was trying to harm the feline. Angrily, she asked Jane what she thought she was doing. Jane quickly explained herself and won Regina over. Regina helped her rescue this cat and many others later on. When Jane's health began to fail because of a fatal illness, Megan, Kate, and Regina began to shoulder the burden of what had become a shelter.

Regina soon began to take a leading role and became the president of the Whiskers Animal Benevolent League. In 1987 she unsuccessfully attempted to form a board of directors to give Whiskers more credibility and expand the shelter's activities. Meanwhile, the apartment in which the cats were housed was in extreme disrepair. The landlord was not willing to make repairs and Whiskers did not have the money to do so. Regina and her associates moved the shelter to the first-floor apartment of a house Kate's father owned. Although they originally saw this arrangement as temporary while they looked for an alternative site, the shelter remained there from 1988 until 2000, which included the entire period of this study–from 1996 until the move to the new shelter site in March 2000.

During Regina's tenure as president the essential philosophy of the shelter emerged and has not changed substantially since that time. From the beginning the goal was to take in stray, abandoned, and abused cats and to maintain them until homes could be found for them. When Whiskers had enough space, they also admitted cats from homes whose owners could no longer care for them. Whiskers was (and continues to be) a "no-kill" shelter, which meant that if homes were not found for the cats, they could stay at the shelter indefinitely. It also meant that euthanasia was considered only when it was the sole humane alternative available, as in the case of terminally ill cats with painful conditions. The shelter was also committed to spaying and neutering to control the cat population.

Everyone associated with the shelter strongly believed that animals did not exist for the convenience of humans.

Rather, they exist for themselves and should have rights that guarantee humane treatment. These beliefs led to the idea that every life was worth saving and that no expense should be spared to save the feline lives in the shelter's care. Shelter volunteers always caged newly arrived cats to ensure the health and order of the shelter. They recognized, however, that as beings who existed for themselves, the cats had a right to some freedom to associate with others and run their own lives. Thus, the healthy cats were always allowed to roam free at the shelter after a period of socialization. Finally, since the cats were valued as individuals, all cats were given names, and photos of all residents were posted so names and faces could be matched.

During Regina's presidency an organizational structure began to emerge. By 1990, she had succeeded in forming a board of directors and had created some regular positions such as volunteer coordinator, fund-raising coordinator, and vet transport volunteers. She also attempted to find a veterinarian who would visit the shelter regularly.

In spite of these changes, serious crises developed that led to Regina's resignation and the formation of a stronger organization. A central problem was the absence of any clear policy on the number of cats that could be admitted to the shelter with its very limited physical and financial resources. When Regina admitted some 50 cats from the home of a collector,[2] the shelter almost collapsed under the burden. The cats were interbred, many were ill, and most were too undersocialized to be immediately adoptable. Megan, Kate, and a relatively new volunteer, Betty, accepted Regina's resignation in July 1991. Megan and Betty became copresidents and met with the board to reorganize the shelter.

The new organization created volunteer positions for major functions and established a structure of responsibility and decision-making between the board, the president, and rank-and-file volunteers. The copresidents' job was pared down to a more manageable level. They were responsible for admissions, adoptions, and initial follow-up calls. The copresidents and board set long- and short-term goals, with debt reduction as the most important short-term goal. To that end, in 1992 the shelter was incorporated as a nonprofit organization under the name Whiskers, Inc. The shelter acquired a veterinary service and cats were tested for, and vaccinated against, major communicable diseases such as feline leukemia. The officers and board created an infirmary and tightened hygiene rules to control the spread of illness. For the first time they set a limit on the number of cats the shelter could hold and established a waiting list for cats they could not admit right away because of space limitations. Copresidents Megan and Betty sent the following memo to volunteers:

> We are currently between $7000 and $8000 in debt, mostly to our vets. That is one reason why we are constantly asking you to sell this or that or go on a cruise [on the Hudson River as a fundraiser] or come to a garage sale.
>
> We have a limit of 65 cats in the shelter at any one time (give or take one or two). We have found out the hard way that any more than that creates stress, which in turn creates illness. So, when you find a stray, we may not always be able to help right away, but we do try to give priority to requests from volunteers.
>
> We have been given new discounted rates at ______Vet by Dr. ________. She is also willing to sup-

> ply us with whatever medications we need at great prices. She is an excellent vet, and a very nice person as well, so if you are looking for a vet for your own pets, give her a try.[3]

Although the limit of 65 cats in this small shelter was still very high, and although this limit was exceeded at times by 10 to 15 cats, the new administration was moving toward more realistic goals. At the same time that the philosophy and mission remained unchanged, the administration developed policies that took into account the shelter's limited resources. As a consequence, Whiskers has thrived and in March 2000 was able to move into a new shelter building purchased with funds raised during the new administration.

Running the Shelter

During our study, shelter philosophy remained as described above and the copresidents and a board of directors established shelter policy. They determined such things as which cats met the criteria for acceptance into the shelter, the maximum number of cats the shelter could hold, and adoption screening and follow-up procedures. Setting a limit on the shelter population was critical because there were always more cats in need than the shelter could accept, and there was always a waiting list. The shelter took in a few cats from homes of owners who legitimately could no longer care for them, but admissions were primarily reserved for stray and abandoned cats. Owners who called wanting to "get rid of" their cat (or cats) were politely but firmly rejected. The volunteers took turns dealing with these unpleasant "dump" calls as they were described. Understandably, the shelter did not publicize its address in order to prevent people from simply leaving unwanted cats on the doorstep–an event that, nevertheless, occurred with some regularity.

The shelter officers took the screening of potential adopters very seriously. Having rescued and rehabilitated many of the cats, they wanted to assure good homes for them. Potential adopters had to fill out a questionnaire and submit to an interview. The shelter officers asked prospective adopters about previous pets and what happened to them, for example. For renters, they wanted assurances that their landlord had granted permission for pets. For roommates, they wanted to know who would take the cat when the household broke up. They wanted to know about allergies in the household and whether there were small children (no kittens were given to households with children under five years old). Adopters had to pay a fee and sign a contract in which they promised to spay or neuter the cat if this had not been done by the shelter, keep the cat indoors, and, most importantly, return the cat to the shelter if they were unable to care for him or her.

The shelter had a veterinary service nearby that cared for all the cats. Given the philosophy of the shelter that each cat's life was precious, this was by far the major expense the shelter faced. The volunteers obtained food and litter mainly from local supermarkets that allowed the shelter to "salvage" supplies that were still wholesome but too damaged to be sold. They paid for shelter maintenance and veterinary services out of income from adoption fees, individual donations, and fund-raising events that the shelter held each year.

All positions at the shelter were held by volunteers. The major positions in the organization besides the presidents and board were the health officer; volunteer coordinator;

cleaner/feeders, who came in twice a day to feed and clean up after the animals; and hugger/lovers, who came in explicitly to give attention to and socialize the animals. In addition, the most closely involved volunteers were responsible for administering prescribed medications, arranging adoptions, following up on adoptions, fielding calls from people who wanted to place cats at the shelter, arranging public events for the shelter, and picking up and storing supplies. Although a few of the volunteers were students or retired people, most of these extraordinary individuals had full-time jobs, and many had families and were involved in other community activities.

During our observations, the shelter occupied the first floor of a two-family house in a residential section of Albany. There were five small rooms plus an enclosed back porch that housed the many litter pans. One of the rooms was maintained as the "isolation" room. All new cats coming into the shelter were caged in this room until they had been medically examined and declared "safe" to enter the general population. A second room served as the "infirmary," where sick and injured cats were treated and restored to health. Both of these rooms were closed off to the general population of cats, who were free to roam throughout the rest of the apartment. Two of the remaining three rooms provided the main "living space" for the free cats and those who were held in cages while they transitioned from the isolation room to the general population. The kitchen was the third open room and served as the feeding area and laundry room. We note two important aspects of the open rooms. First, no space was denied to the cats. This meant that whether you were sitting, filling out forms, dishing out food, or cleaning, you were likely to be nose-to-nose with one or more of the cats. In other words, the cats had maximum opportunity to engage the human members of the shelter community as well as relate among themselves. Second, these rooms were all quite small, averaging 10 by 12 feet. With anywhere from 50 to 60 cats roaming free, this was a very crowded situation.

Over the next several chapters, we describe the nature of the relationships that developed between the humans and the cats and among the cats themselves. These highly positive relationships we believe derived, in part, from the policies and cultural practices that were part of the formal organization of the shelter. For instance, the shelter provided safety by removing cats from abusive and dangerous situations. It spayed or neutered all cats who came into the shelter, not only to control population, but also to eliminate sexual competition. It provided abundant food for the cats, many of whom were abandoned or feral and had to struggle to find and protect their food. The shelter emphasized health care to protect the cats' lives and also to eliminate physically painful or annoying conditions that might have evoked hostility to others. The volunteers introduced new cats into the shelter community gradually by first caging them in the open rooms so they could get used to the other free cats. By following these cultural practices, the shelter removed many of the conditions that otherwise would have evoked fear and hostility in the cats and laid the groundwork for more positive emotions such as affection and friendship. These practices also provided the consistent character of the shelter setting.

Shelter Residents

We can best describe the shelter as an artificially created cat colony with human attendants oriented toward interpreting and responding to the wishes of the cats. This unique setting, neither natural colony nor traditional household, provided an excellent opportunity to advance the ethnographic study of human–animal and animal–animal relationships. Although many conditions for the cats remained unchanged over time, there were also several variables that we focused on to allow us to observe the ways in which the cats adapted to this setting.

Because most cats were not caged, the shelter provided a setting in which the cats were free to make choices. They were free to associate with other cats or to be alone if they could find a spot for themselves. They could interact with humans or not as they chose, although volunteers tried to interest feral[4] cats in interacting with them in the hope that the cats would become adoptable. The cats were free to express themselves in many ways with one another and with the volunteers. The only rule was that the cats were not allowed to fight with one another or attack humans. Rarely cats had to be caged for "time outs" because they picked on more timid cats.

The shelter provided many different kinds of food, allowing the cats choices. If a cat expressed a preference for a food that was not normally provided, the volunteers tried to acquire it. Whiskers even took into consideration the cats' adoption preferences and tried to place them in compatible settings. In addition, the volunteers tried to adopt friends into the same home.

How long cats stayed in the shelter was another important factor affecting the shelter community. Because Whiskers was a "no-kill" shelter, many of the cats, particularly the feral cats, were long-term residents. This allowed us to study the interaction patterns between the humans and cats and among

Monte and Corinne find a little privacy away from the crowd.

the cats themselves over a long time period. We could observe the emergence and maintenance of culture and social structure within the shelter and develop intimate knowledge of individual cats. We believed that only by observing particular cats over long periods could we avoid the tendency of traditional behaviorists to see animals as simply representatives of a species, rather than as individuals.

Whether cats were friendly or feral affected the relationships that developed between the cats and the humans and the formation and maintenance of the cat community at the shelter. We were also able to observe sex differences in cat behavior. Although the impact of this variable was greatly reduced because all the cats were spayed or neutered, sex-typical behavior was by no means eliminated at the shelter. We also observed the effects of age, particularly as the cats became elderly. Because kittens are so vulnerable to disease, the volunteers did not keep them at the shelter but instead put them in foster homes. Kinship groupings such as siblings did occur at the shelter and we took note of them, but they were rare and had little impact on the community. Finally, we could observe the effects of cats entering the community and cats leaving the community either through adoption or death. Both of these events represented potential disruptions to the shelter community.

Knowing the Other

To achieve insight, or intersubjectivity, into the inner lives of others, we took the ethnographic approach of participant observation in which, as Arnold Arluke and Clinton Sanders note, "[we try] to grasp the meanings of the subjects' behavior by seeing things from their point of view."[5] According to Burke Forrest,[6] the participant observer method has evolved over time. Researchers prior to the 1960s were concerned about the objectivity of the participant observer and emphasized his or her need to guard against an overinvolvement that might adversely affect judgment. Consider William F. Whyte's statement regarding his study of the corner boys in *Street Corner Society:* "I tried to avoid influencing the group because I wanted to study the situation as unaffected by my presence as possible."[7] These earlier researchers achieved intersubjectivity primarily through informants such as "Doc" in *Street Corner Society*. In these cases, the informant relays his or her intersubjectivity and understood meanings to the researcher.[8]

Since the 1960s, a new model of participant observation has emerged, which Forrest calls "apprentice-participation."[9] Patricia and Peter Adler refer to it as the "complete membership role."[10] Here the researcher "becomes the subject" by developing close relationships with subjects and fully participating in their world. Only by becoming a "native"[11] can one hope to understand subjective reality. In this model, then, the investigator achieves intersubjectivity directly through his or her own experience, rather than through informants. Forrest and the Adlers were, of course, addressing the study of humans. For us, the question then became how we could achieve intersubjectivity with the nonhuman subjects in the Whiskers community, given our species differences and lack of a common language. Here we agreed with Arluke and Sanders[12] that there were many indications that the "otherness" of nonhuman animals is not impenetrable and that humans and animals can achieve "operative understandings" that not only make routine interactions possible but also provide insights into the animal mind.

Cats seem to interpret many human gestures in the same way. Thus, cats usually interpret patting your leg when sitting down as an invitation to sit in your lap. Beckoning is recognized as "come over here." Many cats quickly learn to participate in a game of peek-a-boo and will alternately look for you and hide themselves. Whether these responses are because of prior experience with humans is unclear. Cats also create signals for humans to learn that then become part of future interactions. Our cat Nicky wrapped his tail around our legs whenever he wanted food. We soon learned the meaning of this gesture and could address his need. Thus, cats, at least, are not passive recipients of human attention. They greet you and make demands based on successful past interactions. When we misinterpret them, they often correct us. For instance, some cats will slap you if you stop petting them to indicate they want you to continue. Others will try to move your hand to where they want to be scratched or vocalize until you give them a preferred food. Their interactions with one another are also often understandable because of such overlap in human–animal expression of emotion and intent. For all these reasons it can be counterproductive to unduly emphasize the differences between us and domestic animals with whom we have such a long history of communication.

We are not helpless in the face of uncertainty in interspecies communication. Among the things we can observe are the choices animals make under different conditions. Some experimenters have done this by providing specific choices for animals and then measuring their responses. Marian Dawkins, for example, conducted studies of factory-farm egg-laying hens, who were crowded into tiny cages with sloping wire floors. She wanted to see whether these hens were suffering, which critics of factory farming believed. Her approach was to "ask the animals" by creating conditions in which the hens could reveal whether they were suffering by their choice of preferred living conditions.[13] Dawkins and other researchers found that certain conditions seemed to "matter" to the hens because they would choose to overcome obstacles to get access to them. These included having litter to scratch and having nest boxes in which to lay their eggs. Dawkins concluded that the absence of these conditions in factory farms produced suffering in the hens.[14]

The studies described by Dawkins were essentially experimental, but we argue that the ethnographic method also allows us to observe the choices animals make but in more natural settings.

Second, we can use the ethnographic method to observe the same animals over time. This enables us to revise and refine our interpretations.

Third, unlike using the laboratory, as researchers in a more natural setting we can observe animals in a nonthreatening situation in which they can relax and be themselves.

Fourth, there may be other humans in these natural settings, such as the shelter volunteers in our study, with whom we can check our interpretations.

Fifth, the ethnographic method allows the time to learn animal gestures, expressions, and sounds that we can use in many ways to further our understanding. Roger Tabor,[15] for instance, used such knowledge to behave like the feral cats in his study to gain greater acceptance among them.

Finally, ethnographic research usually involves collecting both verbal and nonverbal data. Many ethnographers stress nonverbal communication as a major contribution of ethnography that distinguishes it from quantitative studies that omit nonverbal data. They argue that the majority of human communication is, in fact, nonverbal and, thus, to leave out such data is to distort the meaning of human interaction.[16] Such data are even more central and critical to the study of animals. Thus, using a method that incorporates nonverbal communication furthers our understanding of interspecies communication as well as communication among animals themselves.

In summary, we believe that, at least regarding humans and domestic animals, there are "operative understandings" born of long association and evolutionary similarity that allow us to interact routinely in a manner that is mutually satisfying. These understandings work well for all practical purposes and can be enhanced and improved upon through the use of the ethnographic method. We are most likely to go wrong not in our failure to understand the everyday material and social needs of animals but in our failure to understand what may be their "higher needs" for such things as variety, aesthetic enjoyment,[17] and association with kin and friends.

Whiskers and Us

As noted in the Preface, we first became acquainted with the Whiskers shelter in 1989 when we found an abandoned mother cat and her kittens on campus. We supported the unique environment of the shelter, so we became increasingly involved, first as volunteers and later as board members. Although we were not active members of the shelter community when we began our study, we were able to easily resume membership based on our prior involvement.[18] That is, we reentered a familiar situation in which our membership was still recognized and accepted by the other shelter volunteers. Our prior experience as cleaner/feeders in the shelter, as well as our membership in a multi-cat household, also eased our way into the cat community of the shelter. That is, we knew how to "behave among cats."

We enhanced our fieldwork role in one other important respect. As we indicated above, there were many roles at the shelter. As long as you were committed to the goals and philosophy of the shelter, a role could be found for you. When we reentered the shelter, we did not resume our former roles; rather, we created a new role, that of "cat researchers."

The shelter officers and other volunteers readily accepted us in this role not only because of our past contributions to the shelter, but also because they knew that we believed the goals of the shelter always came first. When emergencies arose or the shelter was short-handed, we put down our notebooks and pitched in. By way of illustration, we offer the story of Dylan, which is taken from Jan's field notes:

> It was Sunday and we arrived at the shelter to find a woman with four children outside holding a large cardboard box containing a tiny, black kitten lying on a dirty towel. She said that her daughter found the kitten and they tried to feed her, but she wouldn't eat or drink. We could see that the family was very poor and the girl was crying. We went inside to see if Betty was there, but, as luck would have it, she was at a wedding that day. Nancy, Laura, Alice, and her boyfriend were there. We explained the situation to them and decided that we would call Megan while Nancy brought the kitten inside. The shelter had no account at the Emergency Animal Clinic that was always open at night and on weekends, so Megan said that she would try to find an on-call doctor at the regular veterinary service. Meanwhile, we all gathered round the kitten. Marquis, the shelter's feline security guard, came over and Alice's boyfriend lifted him up so he could see as well. The kitten was limp and cold and Nancy and Linda did not think she could be saved, but we gave her some substitute mother's milk from a syringe and she kept trying to open her mouth to get some. Megan called back and said that she could not find an on-call doctor, so we

should prepare a cage for the kitten and she would get her to the hospital in the morning. We knew the kitten would not live until morning without medical attention. So, I said, "Let's take her to the emergency clinic on our own." Steve agreed, saying, "We have to try to save her." Alice and her boyfriend had placed a small cloth over the kitten for warmth and they agreed also. So, we took off with the kitten. I sat in the back of the car with her while Steve tried to find the fastest route to the Emergency Clinic, which was no easy matter from the shelter. I kept my hands on the kitten to keep her warm and watched for signs of life. Every so often she made a little movement or sound. I didn't dare pick her up as she was so incredibly fragile. It was almost as if there was nothing inside the fur. She was completely limp. It seemed to take forever to get to the hospital.

We finally got there and Steve ran inside with the kitten while I closed the car. He told the receptionists we had a dying kitten and they scooped her up and brought her to the doctor while we filled out forms. After a while the doctor emerged and told us they had to put a catheter directly into the bone to deliver fluid immediately to the kitten who was 10% dehydrated, which was consistent with death. They also had to get her warm as her temperature had dropped to 93 degrees (a cat's normal temperature would be about 101 degrees). With that preparation, the doctor took us in to see her and she was a sight. Lying on a heating pad and surrounded with rubber gloves that had been filled with warm water she was trying to hold her little head up. They estimated her age as 4–6 weeks. She had goop in her eyes from medication and was weak as a dishrag. The doctor said the next two hours would be critical and we should call back at that time. But we felt so elated to see that tiny kitty alive such cautions meant nothing to us. We were just glad we had taken the chance.

We went home and called the hospital at 9 and 12 midnight. She was still alive. At this point we wanted her to have a name and asked the staff to call her Stephanie. Throughout the evening we were on the phone with Megan, Betty, and Harriet (the Whiskers health officer), all of whom were clearly happy we had done what we had done and would take over the care of the kitten after this. So, at 5:30 A.M. we went to the Emergency Clinic to pay the bill and pick up Stephanie to take her to Whiskers's veterinary clinic. When we arrived, it occurred to us that we had been assuming the kitten was a "she" all along without any verification and asked the staff to check the sex of the kitten. Well, it turned out that little Stephanie was a male. We called him Dylan. On the trip to the clinic, I opened the carrier to pet him and, sick as he was, the little guy tried to look up at me and relate to me. I was putty!

Every day Dylan got a little stronger though he would not eat on his own. On Wednesday, I went to see him. He was sitting up and immediately gave me a silent meow in greeting and came to the front of the cage to see me. He played with the tail of the toy mouse I brought him. He seemed to be doing great though the doctors kept warning me that there might be some underlying reason for his failure to thrive

> that we didn't know about and that he wasn't out of the woods yet.
>
> Harriet kept us informed of Dylan's progress. He was eating on his own in two weeks. One of the veterinary technicians brought him home when he no longer needed continuous hospitalization but was not yet ready for a foster home. He seemed well on his way to recovery. Then, one night, suddenly and mysteriously, he died. None of us had the heart for an autopsy. The tech buried him on her property in the country.

The incident was heartbreaking and at the same time was one of the sources of our solidarity with the Whiskers officers and other volunteers. We had gone the extra mile to save a life. In the next chapter, it will become clear that the volunteers saw this as one of the defining characteristics of Whiskers.

In many ways, our research was closest to the apprentice model in our relationship to the volunteers at the shelter. That is, it was more participatory than simply observational. But we also availed ourselves of opportunities to draw on the insights of informants by taking down the comments of volunteers in our field notes and interviewing volunteers at the end of the data collection phase at the shelter. We were also highly participatory with the cats, allowing them to treat us as any other cat or volunteer. We butted heads with them, touched noses, talked to them, let them sit in our laps, and played with them. In addition, we utilized whatever records were available that would give insight into both the cats and the Whiskers organization.

We began our formal observations of Whiskers in the spring of 1996 and ended the formal data collection in March 2000 when Whiskers bought and moved to a house that was to become the new shelter site. We had more than sufficient data by that time, and we feared that changes in the characteristics of the new shelter might alter both human–cat and cat–cat relationships. During the study, we observed the evening shifts at the shelter every Wednesday and Sunday evening. We also observed all evening shifts several times during the study. We began observing during morning shifts, but stopped when we soon learned that this shift was brief and task focused, as volunteers needed to get to their full-time jobs.

We took field notes at the time of the visit and elaborated afterward if necessary. Since we were not observing the relationships among the volunteers, but between the volunteers and the cats and the cats with one another, no one was bothered by our note-taking. We wrote down what volunteers said to one another or to us if it seemed to illustrate something about the nature of their relationship to the cats or to the Whiskers organization. We did not record personal conversations. We recorded everything the cats did in relation to one another, the volunteers, or us. This included gestures, vocalizations, movements, and facial/body expressions.

Since the cats' ability to make choices was a major aspect of our investigation, we focused on the free cats, including their relationship to the caged cats undergoing socialization in the public rooms. We did not observe the caged cats in the infirmary or isolation rooms. We also concentrated on the recurring situations in shelter life that would be most revealing of social relationships between humans and cats and among the cats themselves. Thus, this was a study of the social conduct of cats with humans and cats with one another in a cat community in which sexual competition and competition for survival were not factors.

The Framework of the Study

Martyn Hammersley[19] suggested that ethnographic research emphasizes "theoretical description," of which there are several interpretations in the literature. Some social scientists see ethnographic descriptions as applications of existing theory to their current research to elaborate and refine the theory or extend it to a new area of inquiry. In their view, the role of theory in ethnographic research would be to help them move from the "meaning particular social actions have for the actors whose actions they are" to "what the knowledge thus attained demonstrates about the society in which it is found and, beyond that, about social life as such."[20] Our study is essentially in this tradition. We have previously applied the theoretical perspective of symbolic interaction to human–feline relationships, redirecting it and elaborating on it in the process. We continue to apply this modified perspective to human–feline and feline–feline relationships.

As Mary Midgley noted, "The attempt to make pre-programming account for everything has only been made to look plausible by constant misdescription–by abstract, highly simplified accounts of what creatures do, which are repeatedly shown up as inadequate when anybody takes the trouble to observe them longer and more carefully."[21] Thus, we came to believe from the evidence in our first study that humans and cats, and cats and cats, engage in symbolic interaction and that cats possess at least a rudimentary ability to take the role of the other and use this understanding to shape their interactions to achieve both practical and social goals. Rather than being imprisoned in the present, they have a sense of both past and future, and these understandings do not depend on the existence of human-type language. We believe that for cats, symbolic interaction is fundamentally rooted in a combination of keen observational skills and emotional bonds and intuition. These enable the cats to develop a common definition of the situation and construct shared meanings with one another and with humans.

If cats can engage in such symbolic interaction, they will, given time, produce elements of culture or social organization such as norms, roles, and sanctions. That is, a group of cats over time in the same setting will produce a web of socially transmitted behaviors that constitute that group's solutions to its problems. As Elizabeth Marshall Thomas argues, "Animals most certainly do have culture. We fail to realize this for no better reason than that our experience with populations of wild animals is so severely limited we are not often in a position to see much evidence of culture. Worse yet, we have been conditioned to believe that if we have seen one group of elephants . . . we have seen them all, so we don't even search . . . for cultural differences."[22] She continues, "Culture in cats . . . is less obvious, yet it is there all the same."[23] We also thought that the social organization of the shelter would be the outcome of the interaction among the cats, and between the volunteers and the cats, in which the values and expectations of the cats would be taken into account. Thus, we believed that many of the interaction patterns and cultural elements we observed would be unique to the shelter, while others would be found in all cat colonies.

We combined our observations from our earliest visits to the shelter with our theoretical knowledge to determine the situations that would best reveal elements of symbolic interaction and normative patterns among the cats. These included feeding and eating routines, resting and sleeping arrangements, exchanges of grooming and affection, and other friendship association patterns such as those involved in play. The best situations for the study of human–cat relationships would be feeding and eating routines, medication routines, affection and grooming sessions, and adoption sessions. We were guided by these ideas, but our observations did not constitute a "reproduction" of shelter life.[24] The shelter was a highly complex social structure, and we did not systematically observe everything. Rather, we observed the free population, the cats undergoing socialization, and the shelter routines that were most likely to provide information relevant to our framework. We can now go on to open the door to Whiskers and witness the cats and volunteers together as they carry out the busy routines of shelter life.

3

The Human–Cat Connection

HERE WE FOCUS on the volunteers at Whiskers referred to as cleaner/feeders, who had the primary responsibility of caring for the daily needs of the shelter cats. Each cleaner/feeder generally committed two to three hours one morning or evening a week to the job. We centered our analysis on several questions: How did these volunteers construct the world they shared with the Whiskers cats? Conversely, since the cats were active participants in the creation of their world, how did they construct the world they shared with the volunteers? How did the volunteers mediate between the demands of the board of directors and their policies and the demands of the cats? What was the impact of this negotiated order on both the volunteers and the shelter cats? To better answer these questions, we supplemented our own observations with interviews of cleaner/feeders from the two shifts we consistently observed throughout the study and a few long-standing volunteers from other shifts. Thus, from an organizational standpoint, we were able to examine the informal structure of the shelter that emerged from the interaction between the volunteers and the cats. We explored the ways in which this negotiated order con-

tributed to the formation of a unique human–cat community, fostered the self-development of the cats, and created a context for the cat culture of the shelter, which we will address in the next chapter.

Becoming a Whiskers Volunteer

Whiskers did not maintain demographic data on volunteers. Nevertheless, our observations suggest that the cleaner/feeders came from every walk of life. They ranged from high school students to retirees, but most were in the labor force. Their occupations ran the gamut from professionals to clerical and blue-collar workers. Most were female, but there were always a few male cleaner/feeders in the mix. All marital statuses were represented, and those with children sometimes brought them to help out. Many of the cleaner/feeders extended their service beyond their shifts, returning to spend time with the cats, do needed chores, or contribute their work skills to shelter maintenance. A few were handicapped in some way. Almost all were White. The volunteer coordinator periodically created a list of all the volunteers and their functions in the organization. One recent listing included 77 volunteers, who filled the two shifts per day, or 5.5 volunteers per shift.[1]

Since Whiskers was staffed entirely by volunteers, one could not expect to get a great deal in the way of recognition for service. The shelter did give occasional parties for volunteers, admitted them free to fundraisers, and, at one of these annual events, gave awards for outstanding service. However, the volunteer "officials" worked harder for the organization than anyone else and could not be expected to lavish praise on rank and file volunteers.

The cleaner/feeders' motivation can perhaps best be summed up in David Horton Smith's belief that volunteerism is a form of collective action. He says, "The essence of volunteerism is not altruism, but rather the contribution of services, goods, or money to help accomplish some desired end, without substantial coercion or direct remuneration."[2] When we asked, "What made you decide to volunteer at Whiskers?" most of the cleaner/feeders answered in terms that supported this model of volunteerism:

> I've known about the shelter for some time. I like cats and I can't have any more at home. This gives me more interaction with cats and allows me to do good for cats. I thought I would wait till retirement to volunteer, but I couldn't wait.

> I loved animals. I grew up with them. I felt a need to contribute.

> I read about Whiskers in *Metroland.* It was during the holiday season and they were short on food. I called to make a contribution and talked to Halle. She said they really needed volunteers so I signed up.

Even when conflicts arose between officers and other volunteers, their commitment to the cats kept many volunteers on the job. One volunteer, who contemplated leaving after a disagreement with the volunteer coordinator, said that if she left she would miss the cats.

Some Whiskers volunteers received a fairly thorough grounding in procedure, but some received no "formal" training at all:

> I was trained by Samantha, who was volunteer coordinator at that time. Got a packet of information on

procedures. Read the whole thing. Megan was surprised. Started a shift the next week as a cleaner/feeder.

It was a very brief orientation. I was in a group of about ten.

I learned what to do in the infirmary by watching Chelsea. When I started working in the other rooms Megan and Kate were generally there and could answer my questions.

Wasn't trained. Just put to work. Colors were used to differentiate the function of things. Like blue mops were for the litter room, etc. That didn't last long. I wanted to do the litter room because you can't screw that up. No one else wanted to do it.

Supervision was as variable as the training. Theoretically, there was an officer who was responsible for each shift, called a liaison, and a volunteer coordinator, who had overall responsibility for recruiting, training, and assigning volunteers where they were needed. However, these officers were not always present on every shift, and there was little they could do in the way of supervision if a volunteer did not wish to cooperate. As a consequence, vertical communication in the organization was weak and enforcement of board directives and policies tended to be sporadic. The main mechanisms of organizational control were the liaison system described above, standardized training of volunteers by the volunteer coordinator assisted by written material, and the many notes from officers posted everywhere asking for the cooperation of the volunteers in one matter or another. For example:

> Please!!!! Do Not Change the Thermostat Settings. Call Paula with any issues or problems concerning the heating/cooling system. This is important that you do so instead of trying to fix it yourself. Thanks.
>
> Please do not let caged cats out of cages. They're caged for a reason.
>
> DO NOT put towels all over the shelves. They are disease magnets. Also, blankets spread on the floor are just as bad.
>
> Attention All Volunteers. Please close the inside door *before* you open the outside door. This will prevent "escapes."

Every volunteer had a file in which officers placed various notices and information about the organization. From time to time, Whiskers also published a newsletter containing information for volunteers. However, none of these efforts were consistently applied, and cleaner/feeders were well aware of the communication problems. When we asked volunteers what they liked least about Whiskers, one volunteer said:

> Not getting a grip on the organization as a whole. There is not good communication between the board and the cleaners. I don't feel I am a member of an organization. I also sense some dissention between board members. Sometimes, I read in the paper about things that Whiskers is doing and I didn't know about it and I feel I should. If the board meets, they should tell the members what they are discussing and what decisions they have made–tell the members what is going on.

One consequence of the sporadic nature of vertical communication at Whiskers was the potential for a weak organizational identity among volunteers and, though numbers were not available, there was a good deal of volunteer attrition. However, even though many of the volunteers perceived the structure of the organization negatively they were virtually unanimous on what Whiskers meant to them. Sociologists would see this as stemming from an effective organizational culture that bound members to it through the expressive or symbolic content of organizational decisions rather than their instrumental character.[3] We asked the question, "What story comes to mind that most expresses the nature of Whiskers to you?" A sample of their replies follows:

> Francis's story–he was found near dead in the road and some woman in Delmar called Whiskers. She got him into Whiskers through Betty [copresident]. He was so damaged he had to go to Dr. Barnes [an orthopedic surgeon]. The operation cost $1000. He gave them a break.

> Little Edmund in the infirmary. Wasn't getting better. Didn't want to pick him up but had to cause he was so adorable. Couldn't eat. Thought they'd have to put him down. I thought it was the last time I would see him. He was so affectionate even though he was dying. Then they took him to some specialist in Massachusetts and they saved him.

> A couple of stories. The most extreme was the rescue of the Puerto Rican puppy. [A woman who rescues dogs in Puerto Rico and tries to find homes for them in the states managed to contact one of the copresidents who took this sick and malnourished puppy

> and made him healthy and kept him.] Also, the cats from Japan. Helping this woman who had nowhere to go and little old Whiskers in Albany took them. [The two cats from Japan were leukemia positive. Whiskers took them and found homes for them together thanks to a lot of media coverage of the case.] Also, the policy of not putting down FLV and FIV cats but trying to find homes for them instead. All life is worth saving.

Thus, cleaner/feeders perceived the organizational culture of Whiskers as committed to life at all costs. The staff went the extra mile to leave no possibility untried when it came to saving the life of a feline (and an occasional canine). Many volunteers were amazed that this commitment persisted in spite of a weak organizational structure and limited resources. They found it a source of pride that counteracted the deficiencies they otherwise noted. As one volunteer put it, "Things are never where they are supposed to be and people don't do what they are supposed to do. But volunteers can only do what they can do. But the idea that we are all volunteers. It's amazing what we do. They all deserve recognition."

The stories told above represent one of the cultural forms that characterize an organization. Notice that all the stories involved cats. Of all the cleaner/feeders we talked to, only two discussed human relationships, one indicating admiration for the dedication of one of the copresidents and the other positive feelings toward his shift partners. The main orientation and commitment of the cleaner/feeders, then, was to the cats at the shelter. Given the weak vertical communication and sporadic control exercised by officers over the cleaner/feeders, the practical consequence of this situation was that cleaner/feeders experienced the demands of the cats as more real than the demands of the organizational officials. Hence, in this organization, the negotiated order played a large role in shaping the human–cat community of the shelter, and those negotiations included considerable input from the resident cats. Such an organization may be said to be more socially than policy driven.[4]

How the Volunteers Saw the Cats

In this section we argue that, to a large extent, the social order of the shelter emerged out of the interaction between the cleaner/feeders and the cats. This "negotiated order" was as much a product of cat interests as it was human interests, and a number of factors contributed to this situation. First, because most of the cats were not caged, there was a great deal of human–cat interaction. Indeed, being able to interact with the cats was a major motivating factor for many of the volunteers who either extended the length of their shift or came back to the shelter "off duty" to relate to the cats. Margot, a Wednesday evening volunteer, when asked if she had time to relate to the cats, said, "I have to make the time. When I first started they told me the shift would be one to two hours with time for interaction, but a typical shift for me is 4:00 P.M. to 6:30 P.M., and then, if I can, I may stay until 7:30 P.M. or 8:00 P.M. to interact with the kitties. Or I come in on other days when I'm not doing a shift so I can interact with the cats."

Second, the fact that most of the cats were not caged also meant that the cats were able to make choices and make their wishes known to the humans. The cats' choices included such things as where and when to eat, with whom to relate (cats and humans), and where and when to rest and sleep. Finally, the shelter staff enabled the cats to be full participants in the shaping of the shelter society because they accommodated the cats whenever possible. And most of the volunteers we interviewed believed that some measure of understanding was possible between cats and humans. Let us consider, then, some of the ways in which the norms of the shelter represented a negotiated order between cats and volunteers in which the volunteers strove to take the perspective of the cats.

One such pattern we observed took place at feeding time. Volunteers fed the cats twice a day. This involved eight to nine dishes of dry food and water, and six to seven large plates of wet food, which they distributed around the kitchen floor. Throughout our association with the shelter, there were always two or three cats who indicated a preference to be fed separately, which volunteers invariably agreed to. At one point in our observations, Kemet, Danny, and Cimmeria, the oldest cat in the shelter, were receiving special feedings. Consider the following typical observation of an evening feeding session:

> Feeding has begun in the kitchen and there are about 15 cats waiting. Cimmeria is in her usual place on top of the microwave on the food counter. As Alex [a volunteer] opens each can, Cimmeria sniffs it waiting for just the right dish. When she indicates her preference, Alex makes a small plate for her alone. One of the other cats jumps on the counter and Alex gently sets him back on the floor. Kemet, meanwhile, is making it clear that he will only eat in the sink by repeatedly jumping in and licking the empty cans that have been

> placed there. Alex offers Kemet a small dish to himself in the sink. Danny, who has been rubbing against Alex's legs, jumps on top of the large cabinet where food is stored. As Alex prepares a small dish for him, one of the shelter officers who is passing through the kitchen comments: "Danny has most of the shifts trained to give him a special dish of food."

Another example of the volunteers reinforcing cat-initiated choices was reflected in this early entry in our field notes:

> One of the things that has come to our attention is friendships between cats. One such pair is Bibbiana and Lisa; they are always together and everyone at the shelter recognizes this as a special friendship.

Kemet has trained the volunteers to give him a plate at the sink. All photos by Janet Alger unless otherwise indicated.

> Indeed, they say they will only adopt them out as a pair. Bibbi and Lisa share a special spot together in the kitchen under the heating unit.

Shelter volunteers recognized and accepted friendship choices among the cats. When a particular friendship was identified, volunteers let the others know, who greeted the news with pleasure and satisfaction. Not only were the volunteers happy that the cats would have close companionship, but such friendships also served to confirm the positive character of the cat community. Accordingly, when two cats chose each other as friends, the volunteers generally facilitated this friendship in two ways.

First, if the friendship pair had a special "place" where they associated or slept together, the volunteers would try to maintain such places. Bibbiana and Lisa's friendship activity, for example, was centered in an empty litter pan under the heater in the kitchen, so the volunteers always made sure that this pan was available to them. We have more to say about Bibbiana and Lisa and the course of their friendship in Chapter 4. A second way in which the volunteers acted to maintain cat friendships was reflected in shelter policies on adoption. When it was clear that two cats had become close friends, the shelter would only adopt them out as a pair. We observed this in the case of Calvin and Hobbes, two orange cats who came into the shelter together and who were inseparable:

> Betty has an adoption in progress but it doesn't work out. The prospective adopters are two young women who settle on a cat who is afraid of one of them. The only other cat they want is either Calvin or Hobbes. Betty explains that they must go together as they are

brothers who are very close. The girls say they will think about it and leave.

Although siblings might get attached to each other like Calvin and Hobbes, most friendships were between unrelated cats like Bibbiana and Lisa. When we asked specifically about Bibbiana and Lisa, the copresidents assured us that they would only be adopted out as a pair.[5]

Accommodating cat choices could also be seen in the case of the "feral shelf," a high, wide shelf that ran the width of the litter room–a closed-in porch at the back of the apartment that housed the shelter. Probably because it was very high and the farthest spot from the public rooms, several of the feral cats chose it as their special spot. The volunteers maintained this chosen spot in two ways. They made the shelf more comfortable by putting down blankets and towels and, although storage space was at an absolute premium in the shelter, the volunteers never considered appropriating this space from the feral cats.

Finally, from time to time cats were granted their "choice" to remain at the shelter and not be adopted out. These were friendly cats who had been adopted several times but who always ended up being returned to the shelter. Their adopters gave various reasons for bringing them back, but they always involved inappropriate behavior. Since the reasons for their inappropriate behavior were usually not clear and since they seemed to be happy at the shelter, the volunteers interpreted the cats' actions as indicating that they wanted to remain at the shelter. So they were taken off the list of adoptable cats. In the next chapter we will discuss Marquis, the shelter mascot and feline security officer and an example of this phenomenon. Emilio, a large white male, seemed also to be heading in this direction, but he was eventually adopted successfully:

> Roberta [a volunteer] asks us if we know why Emilio is back. We tell her what Megan [copresident] told us when we asked the same question. The woman who adopted him said he was very unhappy with her. He wouldn't eat, so she took him to the vet. The vet found nothing wrong with him. He also tried to get out of the house every chance he got. So, she returned him. He seems very happy at the shelter. I had asked Megan if he was an only cat in that situation and could have been lonely. She said he had been previously returned by an adopter because he beat up on her other cats. So, maybe we have another mascot.

Although the volunteers generally accommodated cat choices, occasionally shelter officers interfered in the interest of meeting other goals such as maintaining the health of the cats or conserving resources. In some of these instances, the cleaner/feeders mediated between the cats and the officers. One example of this had to do with what we called "soft places." These were blankets, towels, cat beds, and furniture with cushions, which, when available, were the preferred choices of the cats for sleeping and cuddling. During our observations, we saw a "struggle" between the cleaner/feeders who wanted to provide soft places for the comfort of the cats and the officers who wanted to reduce the number of soft places because of legitimate concerns about hygiene and the spread of disease. The officers were successful in banishing furniture with cushions from the shelter, but when they attempted to ban towels and blankets from the cage tops some volunteers (including ourselves) felt they had to intervene on behalf of the cats. The officers complained that these towels and blankets were not being changed regularly and represented a health hazard. These volunteers argued that the cage tops were important sites for cat socializing (more on this in Chapter 4). When we personally promised to change the cage tops twice a week, the officers rescinded the ban.

Cleaner/feeders and officers also disagreed about the amount and timing of the cat feedings. The shelter relied on donations of food. When supplies occasionally ran low, the cats were put on "short rations," particularly of wet or canned food. Cleaner/feeders sometimes disregarded these posted directives or brought food in themselves.

When we first began our observations, volunteers fed the cats in the kitchen immediately after they arrived for their shift. The cats had been alone for several hours, they were hungry, and they expected to be fed. Once the initial feeding was over, the volunteers picked up the plates, swept and mopped the floor, and put the plates down again with more fresh food.

About a year into our study, the officers instituted a new policy whereby the kitchen floor was to be mopped first, and then the cats would be fed. This policy change was upsetting to some of the volunteers who felt the cats' feedings took priority. One volunteer attempted to soften the policy:

> Jill and Anna arrive at the shelter and go into the infirmary to talk to Megan. Jill emerges and says she will feed the cats, but Megan says she must mop the floor first. Jill complains that feeding is the most important thing–"Isn't that what we're here for? The floor isn't even really dirty." She concludes that she will sweep and mop real fast–just a once-over.

We perceived this policy as upsetting to the cats as well, especially when there was a delay in the mopping of the kitchen floor. Consider our observation of an evening shift under the new policy:

> The kitchen floor still has not been mopped and the loose cats have not been fed (it is 6:00 P.M. now). Several have gathered in the kitchen as the cleaner/feeders open cans of food for the caged cats in the front room. Some of the cats rub each other in their prefeeding rituals only to be disappointed. As we leave we comment on the fact the cats in the kitchen seemed agitated and that Samantha [an officer and cleaner/feeder] seemed not to be aware of this; it seemed out of keeping with the general philosophy of the shelter, which is to cater to the cats.

The above example of a lack of consideration for the cats was rare. Typically, volunteers went out of their way to satisfy the needs and enhance the comforts of the cats. During feeding, for example, if there were shy or feral cats who were reluctant to join in the collective eating, volunteers would place a special dish within easy reach for them. Some volunteers regularly brought in special food treats for the cats, such as chicken or tuna, while others brought in fresh catnip (sometimes referred to jokingly as "recreational drugs").

Most of the volunteers had a favorite cat at the shelter. When they arrived for their shift, they sought out their special friend to pet and see how he or she was doing. The volunteers were also especially attentive to cats who seemed depressed or upset at being in the shelter. Alice, one of the Sunday evening volunteers, took a special interest in Glorianna, an older cat who did not like to leave her cage to mix with the other cats. Consider the following entries in our notes:

> Alice is very attentive to Glorianna as usual and takes her out of her cage to brush her. She holds her and hugs her for several minutes.
>
> Alice sets Glorianna at the window in her basket so she can look out while Alice cleans her cage. When that's done, she gently returns Glorianna to her cage. When Alice is ready to leave, she opens the cage, kisses Glorianna goodbye, and tells her she loves her.

The volunteers' concern for the cats' physical well-being is evident in the following incident regarding the two cages in the front room, which we recorded in our notes. The volunteers were faced with the problem of having to change the blanket on top of the larger cage without forcing the cats who were resting there to jump the seven feet to the floor:

> The cleaner/feeders are cleaning in the front room. They are changing the blankets on top of the cages where some of the ferals sleep. One of the cleaner/feeders explains how they do it. The lower cage top was empty, so they changed the blanket on that one first. Then they pushed the other higher cage over next to the lower one and started pulling the old blanket off the higher cage. The cats who were sleeping on the higher cage either jumped over to the lower cage or onto the shelf on the tall climbing pole nearby. Then they put a clean blanket on the higher cage.

In a variety of ways, then, the volunteers took the perspective of the cats at the shelter and, within certain limits, adjusted their behavior to meet the needs and desires of the cats. Most volunteers interpreted official shelter policies within this framework and executed them in such a way as to cause the least distress to the cats. And, on occasion, the volunteers successfully challenged and changed shelter policies.

How the Cats Saw the Volunteers

In the previous section we saw how the volunteers attempted to understand and take the perspective of the cats in the shelter. Successful interaction, however, requires that both parties be able to take the role of the other. In Chapter 1, we talked about how the caretakers in multi-cat households perceived their cats as being able to put themselves in their shoes. In this section we describe behaviors that suggest that the shelter cats were able to take the role of the volunteers. Our focus here is on the "friendly" cats since they regularly interacted with the volunteers and helped to shape the human–cat community. We will discuss the feral cats, who resisted joining the human–cat community, in a later chapter.

It was clear that, to some degree, the friendly cats saw the volunteers as like themselves because they treated them like other cats in many ways. For example, they nuzzled them, groomed them, butted heads with them, or snuggled up against them. Anyone who has owned a cat is familiar with this behavior. Of course, being treated like another cat took a little getting used to as Jan found out:

> Marquis comes into the dining room and jumps up on the file cabinet to say hello to me. He comes up to my face and butts me. He does this twice. The third time he gives me a little bite. I pull away in surprise. He hisses at me. I think it was my fault. In my sur-

> prise I mistook his bite for a danger and, actually, he was being affectionate.

The cats also took the role of the volunteers when they complied with volunteer expectations, that is, when they did the things that humans wanted them to do, such as greeting volunteers when they arrived at the shelter. There were several friendly cats who regularly greeted volunteers at the door, but it was always exciting when a cat that a volunteer had been working with, and trying to form a relationship with, greeted him or her at the door for the first time. This happened to us first with Lisa (of Bibbi and Lisa) and later with Lil'Guy.

The cats also complied with volunteer expectations when they allowed themselves to be picked up or petted, and were cuddly. Consider this interchange with Willy:

> Willy climbs up on Steve's lap! This is surprising because he is so passive. We wonder if he is responding to the attention that we have been giving him or is just coming out of his depression after being returned for the second time by an adopter. I give Steve a brush to brush him with, as he really likes that. Steve eventually puts him down so he can go do something–carry in litter, I think. I sit down and slap my lap for him to jump up and he does. I pet him a while and he jumps on an adjacent desk and sits there with me.

Taking the role of the volunteers was probably best seen when the cats actively, and sometimes aggressively, sought attention and affection. Several cats, like Marquis, could be counted on to jump in your lap for petting the minute you sat down. Others, like tall and lanky Tatiana, would rub or

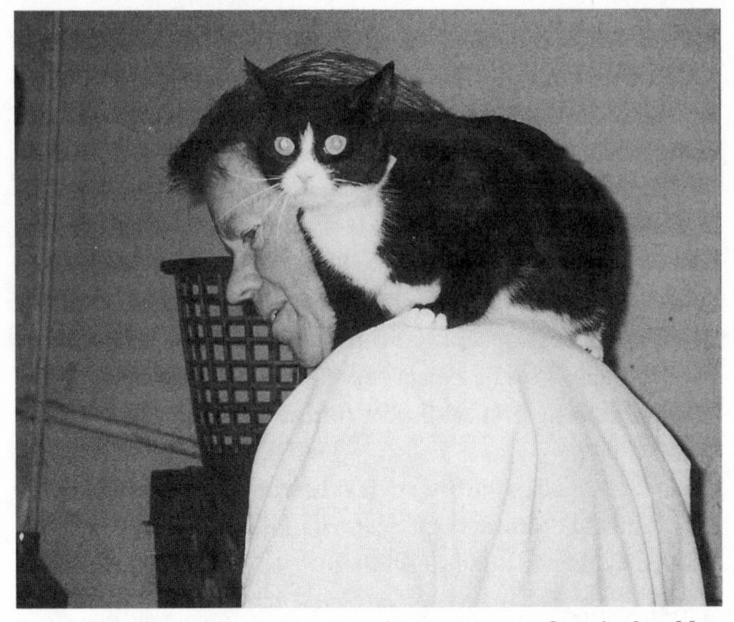

Forget-Me-Not demands attention by jumping on Steve's shoulder.

climb your leg, signaling that she wanted to be picked up. Still others, like Forget-Me-Not, took an even more direct approach; they climbed onto cabinets or cage tops and, as you passed by, leapt onto your shoulders.

Many of the shelter cats sought attention, but some were particularly insistent that the attention, once given, continue. Consider these excerpts regarding Marquis, Amberlee, and Lil'Guy.

> We have not been to the shelter for about three weeks. We start petting the cats and Marquis comes out to greet us. He wants affection so I signal him to jump up on the metal table, which he does. I pet him and he

> stretches up, putting his front paws on my shoulders and nuzzling my neck. When I stop petting he swats my hand. I pull out a chair and sit to give him a lap. He gets in my lap; I pet him; he starts to get excited and nips at my hand. I stand up so he will jump down and as I move away he grabs at my pant leg.
>
> Lil'Guy is in his tree nest by the archway. I pet him and he is very responsive. When I stop he comes down and follows me around. Jan brushes him a bit and so do I. When Jan tries to brush Mystic, Lil'Guy pushes right in and, later, when she is petting Alice, he does the same thing. I go in the litter room to find Amberlee on the warm clothes dryer. I pet her. She gives me little bites and places her paw on my arm to keep me from stopping.

To get attention, the cats would also interfere with the volunteers' activities. Again, anyone who has owned a cat is familiar with this behavior:

> Megan is trying to fill out some forms on the cats. Ophelia is standing, or in this case "sitting," in the way of progress. Megan says Ophelia is always helping with the paperwork. She kisses her.

This "manipulation" of the volunteers was seen earlier in the ability of some cats at the shelter to get special feeding places during mealtimes.

In our study of multi-cat households described earlier, we found that the cats initiated many of the interactions with humans, and the same held true at the shelter. It was mainly through these cat-initiated interactions that the cats made their wishes and preferences known to the cleaner/feeders and, thus, helped to shape the social order of the shelter. This suggests that in the effort to pursue their own goals the cats had at least some ability to take the role of the humans.

The Negotiated Order and the Inner Life of the Shelter Cats

One of the most fundamental contributions of sociology to the understanding of the self in humans is the discovery that the self is social in origin and does not develop without substantial intimate interaction with others. George Herbert Mead demonstrated that

> human group life was the essential condition for the emergence of consciousness, the mind, a world of objects, human beings as organisms possessing selves, and human conduct in the form of constructed acts. He reversed the traditional assumptions underlying philosophical, psychological, and sociological thought to the effect that human beings possess minds and consciousness as original "givens," that they live in worlds of pre-existing and self-constituted objects, that their behavior consists of responses to such objects, and that group life consists of the association of such reacting human organisms.[6]

In short, consciousness and self are not biologically given but must develop through a social process. Specifically, the self arises dialectically through communication with others. Our consciousness of ourselves is based on how we think others see us. Hence, the self is a looking-glass self rooted in the fundamental ability to take the role of the other.[7] It would follow, then, that since the self is not situated within the individual, the development and vigor of the self must wax and wane with the strength of one's social ties. As Robert Bogdan and Steven F. Taylor note in their study of severely disabled people, "We argue in this paper that the definition of a person is to be found in the relationship between the definer and the defined, not determined either by personal characteristics or the abstract meanings attached to the group of which the person is a part."[8]

Although many disciplines have come to recognize the social nature of the self and consciousness in humans, when we study these same things in nonhuman animals, we seem to expect them to be biologically driven.

G. G. Gallup and his colleagues have published the most frequently cited studies of self-recognition in other animals. They studied captive apes who were given mirrors in which they could see their entire bodies. When first presented with the mirror, most species reacted as if the mirror image was that of another ape. When the mirror was left with chimpanzees and orangutans for an extended period of time, however, they learned that the image was of themselves and began to inspect parts of their bodies that they could not normally see. Further, when paint markings were placed on the apes' heads and wrists under anesthesia and they were later allowed to see the head marks in the mirror, they inspected the marks, touching, smelling, and rubbing them, suggesting that they knew the marks did not belong there. The researchers concluded from these behaviors that chimpanzees and orangutans do have a well-integrated sense of self. Other researchers have duplicated these findings for these species but have not been able to duplicate them in the case of gorillas, who are, otherwise, considered as intelligent and as dexterous as chimps and orangutans. They also have been unable to duplicate these studies for lesser apes or monkeys.[9]

To interpret these data, we must first note that, as is common in experiments, very few animals were used in any of these studies: The above-cited study by Suarez and Gallup included four gorillas, two orangutans, and two chimpanzees. It is questionable that we can generalize to an entire species from such a small number of cases.[10] In addition, only one of the two orangutans exhibited the researchers' criterion of self-recognizing behavior.

Although the Suarez and Gallup study included chimpanzees and orangutans, it was intended as a test case for the gorilla. In attempting to explain the gorilla's failure to exhibit self-recognition behavior, the researchers do take account of many nonbiological factors, including the social factor they call "rearing." Yet, there is little doubt that they see their findings as based on the biology of the species:

> In conclusion, it is difficult to explain the lack of evidence for self-recognition in gorillas when they are in a variety of other respects so much like the other great apes.... However, since behavior in the last analysis is nothing more than an expression of neuronal activity, it seems reasonable that species differences in the capacity for self-awareness are ultimately reducible to neurological differences.[11]

Thus, these findings, which are regarded as indicative of the self-recognition abilities of entire species, are based on experiments that include the laboratory behavior of, at best, a few dozen animals and, at worst, just a few animals.

Second, the researchers do not control for wild versus captive birth and upbringing nor do they control for the extensiveness of social ties. Although they do mention that chimps reared in isolation do not indicate self-recognition and they do suggest that the reason one of the study orangutans did not show self-recognition might be because he lived alone, the other animals under study had a variety of backgrounds with different degrees of socialization. The gorillas, they assert, all had extensive exposure to other gorillas, but we do not know how much, what type, or the quality of their social relationships. We conclude, therefore, that social background matters and may be a significant factor in performance on these tests.

We must also point out that these studies are constantly cited by others, as if they demonstrated that all chimpanzees and orangutans can recognize themselves thus furthering the impression that this is a species-wide, biological phenomenon. We are not suggesting here that there are no biological factors at work in self-recognition. However, everything we know about the self suggests that the social factor is central and must be a part of any study of self-recognition.

Oddly enough, an earlier study, in which G. G. Gallup is the lead author, illustrates a greater awareness of this issue than this later one, making it the central concern of the piece. Here, Gallup and his colleagues directly tested the impact of social experience on self-recognition as measured by recognition of one's own image in a mirror. They used six chimpanzees, three wild born and three lab born. The wild-born animals were given companions in captivity and the lab-born animals were isolated. They found that the socially reared animals recognized themselves in the mirror, but the isolated ones did not. After entertaining other possible explanations, they concluded that sociologist Charles H. Cooley's theory of the self provides the best fit with the data:

> To the extent that recognition of one's own reflection in a mirror represents the existence of a self-concept, the feral chimps would seem to have at least a rudimentary concept of self. On the other hand, isolate chimpanzees were characterized by an apparent inability to recognize their reflections. A fundamental and obvious difference between these two groups is in terms of their past history of social encounters. According to Cooley's theory, isolates would not have had the requisite social experiences necessary for the development of a self-concept. . . . Although only suggestive, the present data . . . seem quite compatible with Cooley's theory.[12]

Finally, it is difficult to understand why self-recognition in a mirror has become the *sine qua non* of self-recognition studies. Self-awareness is clearly multifaceted and may be experienced to a degree by many species. For example, most creatures show a preference for themselves. Entomologist May Berenbaum says, "It would be unlikely for an insect to start grooming a neighbor's foot. You wouldn't want to waste energy promoting the well-being of someone else."[13] And most mammals additionally demonstrate an ability to vocalize, show different facial expressions, express emotions, as well as know their place in a social hierarchy. Some show signs of awareness of their own individuality. For these reasons, researchers are beginning to question such heavy reliance on the mirror test. Harvard's Marc Hauser says, "the mirror test is not the be-all and end-all of self-recognition. People have been relying on it too much. What we need are a battery of tests to look at many aspects of self-awareness."[14]

Perhaps we need more than a battery of tests. Perhaps we also need to see animals in social settings other than the laboratory. However that may be, there is nothing in these tests to suggest that the development of the self is not social since it was the more social chimpanzees and orangutans who did best on them. Developmental psychologist Michael Tomasello noted,

> The conflicting evidence about chimp consciousness may reflect the difference between animals raised in the wild and in captivity. He suggests that chimps raised by humans . . . may be more likely to develop a sense of self and possibly an awareness of others. Human babies, he speculates, learn to recognize how others react to them and become self-aware because of the attention they get from adults. Thus, a human raised in isolation might not have the same "consciousness" as the rest of us, whereas chimps raised by people do uncharacteristically well in theory-of-mind experiments.[15]

In addition, these tests simply may not be applicable to all species. Everyone seems to agree that gorillas are as mentally able as chimps and orangutans and, yet, they do not perform as expected. Thus, our best hypothesis remains the sociological one: Self-awareness is developed through intimate interactions with others in which intersubjectivity is realized. As such, it would be as useless to look for the self in the biology of cats as it is to look for the self in the biology of humans. More likely, biology provides a threshold associated with neurological development beyond which aspects of self-awareness become possible, and the degree of self-awareness possible may vary with neurological sophistication. But the realization of this capacity would depend entirely upon the social experiences of the animal involved.

All of our evidence suggests that cats do have self-awareness. We expect that this awareness is strongest in cats who are the most social with humans, other cats, or both. Thus, we look at some aspects of self-awareness that have been identified by researchers and that can be observed in the shelter cat. We then consider the conditions that led to the development of this capacity.

The Social Self in the Shelter Cat

Researchers often use the terms "self-awareness," "self-consciousness," or just "consciousness" interchangeably. Antonio Damasio defines consciousness as "an organism's awareness of its own self and surroundings."[16] However, there is little agreement about the origins or indicators of consciousness. Michel Cabanac, a physiologist, believes that consciousness arose when animals began to experience the emotions of physical pleasure and displeasure. These evolved with amphibians and reptiles such as iguanas, who exhibit a rise in body temperature and heart rate when they are pleased. Thus, Cabanac associates consciousness with emotions that are housed in the more ancient parts of the brain below the cortex and exist in many species.[17]

Psychologist Stephen W. Porges defines the self in terms of acting upon the world and sees the emergence of self-awareness as linked to the evolution of mammals rather than reptiles:

> Mammals, by comparison, are dynamic foragers and explorers of their environment. Regardless of external circumstances, their core body temperature and metabolism are maintained stably by the brain-stem and hypothalamus, thus freeing up neural circuits to permit an active as opposed to reactive stance in the

> world. With the development of the cortex, mammals, particularly primates, gained the ability to engage with others, vocalize, display facial expressions and otherwise show evidence of emotions, all of which ... [count] as aspects of self-awareness.[18]

Cognitive ethologist Donald R. Griffin and biologist Marian Stamp Dawkins emphasize the ability to adapt to novel circumstances as evidence of consciousness.[19] Dawkins offers five indicators of animal consciousness:

1. The complexity of behavior such as the different alarm calls used by vervet monkeys for different kinds of predators.[20]
2. The ability to adapt behavior to variable conditions, which suggests the ability to learn.[21]
3. The ability to learn from others and not merely from one's own experience, which suggests the passing on of information from one animal to another or the development of culture.[22]
4. Behavior that involves choice such as the actions of cock house sparrows, who must decide whether to swoop down after food when a cat is in the vicinity. Researchers have found that their behavior varies with different conditions; the sparrows seemed to weigh alternatives and risks before acting.[23]
5. Cooperative behavior such as the food-sharing of vampire bats as evidence of consciousness in animals.[24]

Sociologists Bogdan and Taylor dealt with some of these issues in their study of the severely disabled. One of the ways in which the nondisabled caregivers constructed their disabled partners was to see them as individuals with unique characteristics. Individuality, or the ability to express oneself in unique ways, would certainly be an indicator of self-awareness. In particular, the disabled people were seen as exhibiting clear preferences and having distinct personalities, feelings and motives, and biographies that explained their nature and behavior at the time.[25]

As mammals, cats display all of the characteristics Porges associates with self-awareness. They interact with their own and other species, vocalize, must act upon the world to survive, and have a large variety of facial (and bodily) expressions. They also show evidence of experiencing emotions, important indices of consciousness noted by both Cabanac and Porges. Researchers can document cats' reactions biologically by looking at variations in heart rate, blood pressure, and hormonal output or behaviorally by examining variations in facial and bodily expressions, vocalizations, and other actions. As we show in Chapters 4 and 5, cats readily adapt their behavior to novel circumstances, which is one of Griffin's and Dawkins's criteria of self-awareness. We also argue in those chapters that their adaptations and negotiated behaviors are complex, that they learn from the experience of others as well as their own, that they make choices, and that they sometimes cooperate.

Here we focus on the aspects of self-awareness identified by Bogdan and Taylor as well as on emotions that several researchers associate with self-awareness. Our earlier discussion about Cooley's theory of the self would lead us to expect that the cleaner/feeders, and some other volunteers who spent a lot of time at the shelter, would play a large role in increasing the self-awareness of the shelter cats because of the attention they gave them. The volunteers would seemingly have the greatest impact on such things as life histories, likes and dislikes, feelings and motives, and personality.

When new cats arrived at the shelter, the volunteers constructed the cats' life histories from the often-little they knew about their past. These biographies started the process of individualization[26] as they were passed on from one volunteer to the next. Some examples follow, and others will appear throughout the study:

> I ask Megan about Ivy. She came from the home of a confused woman who let strays come and go through her open window. The woman asked for help. Whiskers went and took what they could catch–Ivy, another adult, and some kittens. Ivy may be about three years old.
>
> Kate is in the shelter working with three new females, all of whom had been displaced by the fire down the street. One had had kittens who were rescued, but they died from smoke inhalation. Teaser is also from the fire and, apparently, there are a couple more males they haven't caught yet. The owner did not want them back and just let them go because they reminded her of the fire.
>
> Before shift Nancy tells me about a new beautiful white cat in a cage in the living room. She found her and two kittens–a Siamese-looking one and a calico–at the side of a deserted road. She took them to Whiskers at Megan's okay. They all have tattoos that mean they were spayed at a shelter. The kittens are no more than six weeks old and don't look like the white cat.

Notice that these biographies contain physical descriptions of the cats and narratives of the known events in their lives.

But many of these accounts also contain descriptions of character traits such as vivacious or ebullient (see the story of Amberlee's recent history later in this chapter). Although the cats were not aware of the life histories that had been constructed for them, those biographies did affect their interactions with shelter volunteers who reacted to the cats in terms of their past lives and attributed characteristics.

At the same time that life histories were being constructed and passed on, the new Whiskers cat was given a name. Generally, the person who brought the cat in was allowed to name him or her, and cats who were related to each other were given names with the same first letter. Many of the cleaner/feeders made a real effort to learn the names of the cats using the posted photos to make identifications. Many of the cats and most of the long-term residents knew and responded to their names. Remarkably, this included the feral cats who "answered" to their names though they would not permit physical contact. Philip, for instance, would stop whatever he was doing and look straight at you if you called his name.

The shelter cats exhibited strong likes and dislikes, which the cleaner/feeders observed, passed on, and often encouraged. This is clearly shown in the earlier quoted field note in which the volunteer Alex waited for Cimmeria to express her food preference and then gave her a plate of that food. She also prepared a separate plate for Danny and fed him on the cabinet where he was waiting for the special offering. We often learned about the cats' preferences from the volunteers:

> Bibbi now swats two cats away from a plate and eats. The volunteer says she loves Tender Vittles and was waiting to see if she would be given some before eating the wet food.... The volunteer indicated that the shyer cats wait for the second wave of food when the

human and animal activity level dies down before they come out to eat. Penny Royal (brought in as a mother with kittens 6 months ago) sits on the microwave. The volunteer says she is friendly with people but she is aggressive with other cats.

As can be seen in our study of multi-cat households described in Chapter 1, cats show preferences for all major features of their lives. At the shelter, they indicated strong food preferences and preferences for the company of particular people or cats. They also favored particular sleeping spots, games, and toys. Preferences, such as these, are really choices that the cats are making among various possibilities.

The shelter cats also exhibited a wide range of emotions and the desire to communicate those emotions to others. Biologists and sociologists define emotions from quite different perspectives. Biologist Marc Bekoff defines emotions as "psychoneural processes that express themselves as mood." He notes that they relate to feelings such as "love, hate, fear, joy, happiness, grief, despair, empathy, jealousy, anger, relief, disgust."[27] Sociologist E. D. McCarthy defines them as "collective ways of acting and being; they are 'cultural acquisitions' determined by the circumstances and concepts of a particular culture, community, society."[28] In the context of the shelter, the sociological definition is more useful because it points to the way in which the emotions that dominate a setting are a product of interaction over time and are collective phenomena.

We have seen instances of negative emotions such as fear, anger, irritability, frustration, depression, general unhappiness, and jealousy at the shelter. Fear and anger are not easy to distinguish. Consider the behavior of Black Amber, a new cat who was in one of the socialization cages in the front room. A prospective adopter wanted to interact with her, so Megan released her from the cage. They got along fine, but while he was filling out the adoption papers, we witnessed the following:

> Black Amber remains out of the cage. Her eyes are dilated [a sign of fear and distress]. Barbara comes by and Black Amber attacks her fiercely. Marquis rushes over to break it up, but Anna [human] has just arrived and she breaks it up. Barbara, normally a feisty little cat, surprisingly, got the worst of it.

Cats at the shelter often exhibited irritability when they were newly released from the socialization cages and when the volunteers failed to meet their expectations:

> Barushkin is now out of his cage. He is very irritable, swatting cats who come too near him. Kate seems to like him. She fusses over him. This seems to calm him down, at least for the moment. Marquis is also irritable. He swats Lillia and Popsicle. I think it's because Bill [a volunteer] was holding him and put him down abruptly. He comes to us and clearly wants to be picked up, but Steve has Forget-Me-Not on his shoulder. Then he swats another cat, goes to Anna to be picked up but she now has Forget-Me-Not on her shoulder!

A variety of situations can evoke depression in which cats lose their appetite, appear dull and expressionless, and frequently sleep or huddle in corners. Cats who are ill or are being abused by another cat frequently exhibit these symptoms. Amberlee became depressed when she was returned to the shelter after her second adoption:

> I go into the living room where Megan is cleaning cages and, much to my surprise, Amberlee is in a cage. I ask Megan about it and she says Amberlee has been returned by the people who adopted her because they are moving to Florida! After four to five months. Megan lets her out of the cage. She looks depressed to me. She sits on a pole shelf. I pet her and tell her it will be okay. She turns to me and gives me a long, frustrated-sounding cry. Her eyes lack expression. She used to be so vivacious and ebullient. This was her second home. The first people adopted her as a kitten from Whiskers and then returned her for urinating outside of the litter box. It turned out she had a bladder infection. They had kept her for over a year.

Jealousy is not an entirely negative emotion, and the instances of it we saw at the shelter were really rather amusing. We made friends with Lil'Guy, a feral cat at the shelter. As he began to respond to us, he became very possessive:

> I greet Lil'Guy, who is his affectionate self. I talk to the caged cats and he tries to intervene and distract me. So, I go over to pet him and also Laurel and Zoe. He gets jealous as usual and tries to get between me and the other cats–gently. He even moves to kiss Laurel once I go back to petting him.

> I go to pet Lil'Guy, who is on the carpeted two-story condo in the living room. Another little condo is on top of it with Amberlee in it. I pet Lil'Guy and then, while I'm petting Amberlee, he wiggles his way between us. When I go back after doing the cage tops, he is in the house with Amberlee.

The shelter volunteers dealt with negative emotions in a variety of ways. They particularly discouraged cats from acting out anger. In the example above, Anna immediately broke up an attack by Black Amber on another cat. In the next chapter we show that such attacks almost always involved cats newly released into the free-roaming population or cats who had come from single-cat homes. In all cases the volunteers broke up the fights and either put the attacker back in a cage for continued socialization or put him in a "time out cage" to cool down. If the situation was not serious, as in the case of Barushkin, above, volunteers might simply support the cat through the adjustment period of interacting with such a large number of other cats. Support took the form of extra attention and consideration. The shelter staff also took seriously cases of depression. If they thought a cat was depressed, such as Amberlee, they would give her lots of attention. They might put up notes regarding the situation to encourage volunteers on different shifts to spend extra time with the depressed cat. They might also make special efforts to get the cat adopted by taking him or her to adoption clinics and pointing out the cat to prospective adopters at the shelter as in the story of Nathan in Chapter 6. These cultural practices associated with both the formal and informal structure helped to shape the emotional culture of the shelter. We can see these practices as a kind of emotional socialization[29] for the cats–aimed at reducing fear and hostility and promoting affection and sociability.

We have seen numerous instances of displays of affection and happiness among the cats and between the cats and the volunteers. We have also witnessed evidence of empathy in the relationships, particularly between the free and caged cats. First we look at affectionate behavior:

> No volunteers in the shelter when we arrive. The cats are mostly sleepy. Marney and Laura are very affectionate with each other. They are rubbing faces, butting heads, and rubbing against each other. Corey walks around greeting several cats with rubs. He and Lotus give each other additional serious rubs and head butts.
>
> Lisa is holding Mr. Kitty who is licking her face. She calls him her buddy. She is one of the Wednesday night cleaner/feeders. Lil'Momma is in Black Orchid's old cage. Black Orchid goes over and chortles and makes like she wants in. Lil'Momma runs to the corner of the cage to her bed in fear. Black Orchid walks away complaining. Bill picks her up and cuddles and nuzzles her. She loves it.

Other examples of affection between volunteers and cats have already been described; recall the exchanges between Alice and Glorianna and between Megan and Ophelia.

Happiness was most often expressed at the shelter after dinner when the hubbub of cleaning and feeding began to die down. At that time, the cats often engaged one another in play, bathing, and napping as well as playing and smooching with the volunteers:

> I play with a string with several cats–Virgo, Lillia, Marquis, Larry, and others. . . . Virgo keeps getting the string in his teeth, making it difficult to play. The play atmosphere is infectious now that Nancy and Alice have finished their work and left. The place is quiet. Sonnet comes over and licks Virgo, then holds him down and licks him more vigorously. Then she

> wrestles with him. He seems to love it. Meeko and O'Connor start wrestling. Even though Meeko is smaller, he is a handful and definitely holds his own. Creamsicle and O'Malley also have these great wrestling, washing, and playing sessions–usually, at this time of day, after the cleaning crew is winding down and things have gotten quieter.
>
> There's a track toy in the dining room. Feral Mandy sits on the cardboard in the middle and wildly plays with the ball around the track. Then she scratches on the cardboard. We've never seen her in such a good mood. We walk over to the cage tops where all is quiet. Everyone is asleep and slightly separate. We go to change the cage top linens and Lexi goes over to Kate and licks her. He sits mashed up against her and watches us. Alice finishes her work on the living room cages and sits down in the living room to socialize with the cats. Tabatha gets in her lap and then Bruno.

Because volunteers did not often reveal emotional neediness at the shelter, we were unable to observe empathy in the relationships between the volunteers and the cats.[30] However, as we reported in Chapter 1, owners perceived that their cats demonstrated empathy toward them and generally reciprocated in the relationship. If we define empathy as an understanding of what another is experiencing and the motivation to respond to it, we have observed many instances of empathy among the cats at the shelter. The incident reported above in which Marquis rushes over to break up a fight might be seen as empathic. In fact, much of his role-related behavior reported in Chapter 4 appears that way. In Chapter 4 we relate an incident in which a new cat whose owner had died arrived at the shelter. He faced the back of the cage and howled until two resident cats came over and calmed him. A few additional examples will suffice here:

> Donald gets out of his cage while Megan cleans it and makes for the kitchen where there is food. Megan gets him back in the cage and he starts crying. Lotus goes over and kisses him through the bars. Bart is in the individual cage in the living room. He is hot for treats. Duncan goes up to the cage rapidly, nose forward and then rubs against the cage a couple of times. He then sits there and looks at Bart and then begins looking around. Then Bart comes up to touch noses with him. Duncan isn't looking at Bart, so Bart puts his paw out toward him to get his attention. Duncan turns and sees him and they touch noses. Then Duncan walks away. Duncan often visits Bart in his cage.

> Kate is working with the newly released feral Shelly. Shelly is very nervous. O'Malley stands next to Shelly watching everything Kate does carefully. Every few seconds he kisses or licks Shelly's head. Is he playing a Marquis-like role?

The responsiveness of the cats to one another at the shelter was very striking and will be reported on at even greater length in the next two chapters. Just as both the formal and informal structure of the organization discouraged negative emotions, they strongly supported and provided opportunities for the emergence of positive emotions. The cleaner/feeders fussed over the cats, often staying beyond their shifts to relate to them. Special volunteers, the hugger/lovers, came in just to socialize and relate to the cats. And Kate, the shelter specialist in socializing feral cats, did extraordinary work with them. Combined with the freedom from want that the shelter provided for these disadvantaged animals, these practices created the framework for an emotional climate that was highly supportive, affectionate, and benign. As the reader will see, this emotional climate allowed the cats to develop a culture and social system that went beyond the level of simple material need satisfaction.

The use of names, the support for cat preferences, the freedom to express all but the most negative emotions, and the freedom to associate with a wide range of residents and volunteers created a complex environment that fostered a high level of individuality among the cats. Insofar as individuality is the ability to express oneself in unique ways, the shelter cats used many nonlanguage resources to communicate with shelter staff and make their wishes and feelings known. They initiated many interactions with volunteers. In addition, they were always trying to understand the volunteers by examining their faces for clues and cocking an ear, anticipating a volunteer's speech. This latter habit was acquired over time as the new feral cats were often frightened of human speech and simply ran when they heard someone speaking nearby.

Many of the cats greeted the volunteers upon arrival with methods as varied as the cats themselves. Some trilled, some reached their heads toward them, others rubbed against them, and still others found high perches from which they could head butt the volunteers. The same variation in style was true when they sought attention, affection, food, or play. Although not the subject of this chapter, the individuality of the shelter cats also emerged in their eating behavior, their relationships with one another, and their mien when they were alone. This individuality was also true in the complex environment of the multi-cat households reported in the first chapter. Along with the striking repertoire of biologically given ways cats have to express themselves, their individuality provides compelling evidence for the self-awareness of cats.

A Unique Community

The human–cat community we have described was a product of the dialectical process of interaction in which both cats and humans played a central role. It was a complex social environment resulting from the interplay of cat demands and preferences, shelter board policies, and the interpretations of the cleaner/feeders who mediated between the two. Hence, the human–cat community was essentially a negotiated order in which both cats and cleaner/feeders benefited. The cats had the freedom to make choices and express preferences, and the cleaner/feeders had the satisfaction of forming relationships with, and meeting the needs of, the cat population. The reciprocal nature of these relationships was similar to that found in the multi-cat households described in Chapter 1. The fact that this human–cat community "worked" gave evidence for the existence of intersubjectivity between the humans and the cats. Without the ability of the humans and cats to develop shared definitions of the situation, it would not have been possible to have 50 to 60 free cats and their human caretakers in four small rooms without conflict and even some danger. This interspecies understanding was aided by emotions that are housed in ancient centers of the brain and are common to all mammals. The cleaner/feeders discouraged the negative emotions of fear, hostility, and depression, and encouraged the positive emotions of affection, happiness, and empathy. The resulting emotional climate enhanced the human–cat relationships beyond what would have been possible by just fulfilling the cats' material needs as in an ordinary shelter. It allowed the cleaner/feeders to fulfill the cats' social needs as well. That they succeeded in this was evidenced by the many interactions initiated by the cats seeking social goals. The emotional climate of the shelter also provided the context for the cat culture that emerged at the shelter and will be the main focus of the next chapter.

Finally, this complex and emotionally rich environment fostered individual self-development, allowing for the emergence of unique and multifaceted personalities among the cats. Specifically, the shelter staff provided the social intimacy that enabled a social self to emerge among the friendly cats. In the next chapter in which the cats are the primary actors, we will address whether the cat community also contributed to the emergence of a social self among the shelter cats.

4

The Social Bonds among the Cats

IN THIS CHAPTER we describe the social structure and culture that emerged among the cats themselves and their impact on the social self of the cats. The fact that we can talk about "animal culture" is the result of major changes in our thinking about animal behavior and societies. Those who held traditional views saw animal behavior and animal social systems as a product of genetics and evolution and thus fixed and immutable for an entire species.[1] Recent evidence refutes this view and reveals animals to be far more flexible, both in specific behavior patterns and social arrangements, than previously thought.[2] It is when this flexibility leads to variations in behavior and social structure within the same species that animal researchers begin to talk about animal culture, that is, learned and shared behavior among animals.

The most common method used to distinguish culture from instinct is to observe the same species of animal in different physical settings to determine if behaviors emerge that are unique to that setting. Researchers in one of the earlier studies using this method looked at a colony of Japanese macaques that had been transported from their native Japan to Oregon, where they encountered a very different physical environment. In Japan, where the colony was spread over a large area, adult males tended to congregate separately from females and their children. When juvenile males got into disputes, it was the nearby mothers who intervened to protect their sons and, thus, male dominance was strongly influenced by the fighting ability of the mothers. Sons with the fiercest mothers grew up to be the dominant males. In Oregon, the macaques occupied a much smaller area, and it was the adult males, who were now close at hand, who moved in to break up conflicts between younger males. The males, then, established dominance by fighting as adults, and the females did not play a role in the dominance system.[3] Another example of this approach to discovering animal culture can be seen in the study of dialect variation within the same species of songbirds. Different populations of sparrows, for example, often develop their own distinctive song patterns.[4]

One of the most recent and extensive documentations of animal culture comes from a group of primatologists who compared their observations of chimpanzees in seven well-established field sites.[5] They "came up with 39 behavior patterns that fit their definition of cultural variation, meaning they were customary in some communities and absent in others, for reasons that could only be explained by learning or imitation."[6] These cultural variations included such things as the tools chimps used to obtain food, the methods they used to attract the attention of other chimps, and whether the chimps engaged in rain dances (slow displays at the start of rain).

A second approach to separating culture from instinct is to establish that certain behaviors are transmitted through social learning. Many researchers, for example, have focused on learning through imitation among primates.[7] And Dawkins has summarized extensive research on rats that indicates that critical information on food safety (i.e., avoiding poisons) learned by one generation is passed on to the next generation.[8] The sharpest critics of claims for animal culture tend to focus on the adequacy of the evidence for cultural transmission.[9] Here, we are inclined to agree with Frans de Waal and others who downplay the importance of mechanisms of transmission. He argues that "the concept of cultural propagation does not specify whether it rests on imitation, teaching or language. The 'culture' label befits any species, such as the chimpanzee, in which one community can readily be distinguished from another by its unique suite of behavioral characteristics."[10]

Several researchers who study domestic cats have documented flexibility in their behavior and social arrangements, depending on the cats' setting and circumstances.[11] Among these researchers, however, only Elizabeth Marshall Thomas identifies these behavioral variations as culture. She defines animal culture as "a web of socially transmitted behaviors," which constitutes the solutions a social grouping of animals has developed for solving its everyday problems.[12] She gives an example about her practice of taking cats from her son who, like she, lived on a farm. Each time she took a new cat from her son, her current cat would make way for the newcomer by moving to one of the outbuildings. This cultural trait developed, she argued, because these cats were all raised in a setting with numerous outbuildings available. This practice ceased, however, when her son moved into a suburban home with no outbuildings. The cats raised in this setting no longer moved out when she took in a new one. Having been raised in a setting with no outbuildings, these cats solved their problems differently. They stayed on in the house and adjusted to the new cats.[13]

Like Thomas's cats, our shelter cats also developed a culture, that is, a web of socially transmitted behaviors that represented their particular adaptation to the conditions of the shelter. As we describe in this chapter, it was a culture in which the roles, norms, and sanctions strongly promoted affection, friendship, and social cohesion and in which the opposing forces of aggression, dominance, and territory were of little social significance. Since these latter three are the factors traditional animal behaviorists have argued structure animal life, our findings represent a major departure from past research. Although we studied one group of cats only, we compared our findings with other studies of cat groupings. In addition, to the extent possible, we attempted to identify the mechanisms whereby their culture was passed from old-timers to newcomers in the shelter. In the previous chapter, we described how the cultural practices of the shelter volunteers facilitated the emphasis on affection, friendship, and social cohesion in a variety of ways. In this chapter we explore the distinctive contribution of the cat culture to the social organization of the shelter.

Affection, Friendship, and Social Cohesion[14]

Both Paul Leyhausen[15] and Roger Tabor[16] note that animal behaviorists have been obsessed with the study of aggression and individual territory, seeing aggression as a means of maintaining ownership of a territory and its food supply.

Leyhausen and Tabor provide evidence that such behavior is often modified in practice to meet different goals under different conditions. Tabor also notes that studies of affection and similar socially cohesive forces among cats are very rare. This is so, even though his own observations found many more instances of affection than aggression among free-living cats. He concludes that, perhaps, aggression is easier to observe than positive actions, which may be more passive–such as sitting quietly together near one another.[17] Tabor does indicate that the cats he observed have a strong sense of personal space. If their space is intruded upon, they will issue a warning. He further notes that cats who are kin tend to have the fewest barriers between them.

Our long association with the shelter has yielded so few instances of serious aggression among the free cats that we would have a short study if that were our focus. Rather, we wish to make a contribution to the understanding of the far more frequent instances of affection, friendship, and social cohesion that we witnessed. We observed cat interactions in three settings. The first was the cage-top resting areas. For most of our study there were two large, multilevel cages in the front room that housed new cats who had been inoculated, spayed or neutered, and declared healthy by the veterinarians. The cats remained in these cages until they indicated a readiness to join shelter life. The wire tops of the cages were covered with blankets to provide resting places for the free population and to allow the caged cats to become accustomed to the close proximity of other cats. The second setting was the various chairs, cat beds, cat shelves, and so on, that were scattered throughout the two front rooms. The third locale was the kitchen in which the cats gathered to eat at feeding times.

The Cage Tops

The following observation of cage-top behavior was typical:

> We arrive at the shelter and find seven cats on top of the big cage. Scamper and Alice are snuggled together on one side and Carly, Jenny, Logan, Philip, and Merlin are snoozing on the other side, all touching one another. Jenny gets up and moves in between Logan and Philip and settles down against both of them. Tess, who was on the shelf near the big cage, jumps on the cage top. None of the others show any reaction. She sniffs Philip and then snuggles in between Philip and Merlin.

There were always cats on top of the cages, often seven or eight to a cage. If that number of cats were equally distributed, there would have been, perhaps, four inches of space around each cat. Instead, they sat entwined and sandwiched together without the slightest regard for personal space. In these positions they washed one another, slept, and cuddled together. Acts of aggression among the cats on the cage tops were rare. Such conviviality is not common to all cats, however, as the new cats were often aggressive. When Susie was first caged, for instance, she rushed to the edge of the cage and hissed, snarled, and swatted at the resident cats as they climbed up.

Over the course of our observations we identified 93 different cats of all sexes, ages, and levels of tameness on the cage tops. Length of residence at the shelter was the only background factor affecting the relaxed atmosphere of the cage tops; most cage-top cats were old-timers.[18] We did observe occasional episodes of hissing between old-timers and newcomers:

Everyone is welcome to snuggle and snooze on the cage tops.

> Tiberius [new cat] tries to climb onto the front room cage top occupied by several cats and Merlin [old timer] hisses in a lazy sort of way, making no effort to get up. Tiberius slows his approach in response and finally settles down on the opposite end. This satisfies Merlin, who yawns, and then everyone settles down and goes back to sleep.

The only other factor affecting the overall sociality of the cage tops was personality. Occasionally, a cat would come into the shelter who was slow to adapt to the peaceful atmosphere. Sid, whom we all called "Sid Vicious" after the punk rock star, was such a cat, but even he came around and eventually left everyone alone most of the time:

> The big cage has Dot, Danny, Scamper, and Sid on top, all separate. LaSalle climbs to the top on the side

> near Sid. When he sees Sid he starts edging over across the back of the cage without climbing up. He climbs up when he's a ways from Sid, who hisses and makes like he's going to go for LaSalle but doesn't follow through. LaSalle jumps down, nevertheless. Merlin goes up to the top and tries to settle next to Sid, who snarls. He settles, nevertheless, and Sid leaves him alone.

Although acceptance of others on the cage tops was sometimes granted grudgingly, it was clearly the norm. During the entire period of our observations, we never saw any cat driven off the cage tops by another cat.

Beds, Chairs, and Sleeping Spots

Cats also rested and slept in the many cat beds, baskets, cat shelves, chairs, and other human purpose surfaces in the shelter. Some beds, baskets, and shelves were designed to fit one cat, while others could easily accommodate two or three cats. Every time we visited the shelter we took note of who was occupying these various sleeping spots and, thus, have a record of the many instances of multiple cats sharing the same bed. Cat friends often slept, rested, or groomed together in a bed meant for one. Sometimes two cats, such as Cedric and Mickey, were involved in this behavior, but we often observed three cats sleeping together in a bed for one. Once, on a cold winter day, we actually saw five cats sleeping in a bed for one. A couple of them could hardly be seen under the others!

Many cats chose to sleep in an occupied bed even if it was a bed for one and there were other beds available. Geraldine, a young short-term resident with a spinal injury, frequently

Best friends Merlin and Corey share a bed.

sought beds occupied by other cats before her death at the end of March 1997. On one occasion, she nuzzled into a single bed with Chelsea, who immediately woke up and sniffed Geraldine, who had soiled herself. Chelsea accepted her presence, which she indicated by going back to sleep, whereupon Geraldine pushed her face into Chelsea's flank and also went to sleep. Some time later she awoke to find Chelsea gone, as Chelsea had decided to get some dinner. Geraldine then left that bed and tried to enter Tara' s bed. Tara swatted her and kept her from settling there.

Many cats, like Chelsea, were very tolerant of varied sleeping partners and seemed to attract other cats like Geraldine, who wanted to cuddle but did not have a special friend of their own. As a consequence, Chelsea, and cats like her, rarely had a bed to themselves for long and, thus, played an important role in shelter solidarity. Only a few, like Tara, would not permit other cats to join them. Still others, like Merlin, resisted sharing their beds at first but then completely gave in to sharing. Merlin, in fact, became what might be called the patriarch of the shared sleeping space.

Friendship Patterns

If friendship is to exist in any species the friends must demonstrate a regular and persistent preference for each other in a variety of ways. At the shelter, the cats indicated such preferences in eight distinct ways. Cats who were friends might regularly greet each other when they met by butting heads or touching noses. They might rub against each other, sometimes in greeting but often just for the sake of doing so. They "hung out" together, which might involve simply sitting together or sleeping next to each other. They might cuddle together in close physical contact. Cats who were friends might regularly groom each other or they might play together. Often, friends would eat together, and, in one case, they maintained a special place of their own together.

Such friendships were common at the shelter. Sometimes they involved two cats, but, often, there were three or more. We judged the degree of friendship either by the number of friendship activities the group engaged in or the exclusiveness of the relationship. Let us begin with the example of Bibbianne and Lisa, who met at the shelter and became close friends. Bibbi was a brown tiger with a lot of white on her underside and legs. Lisa was primarily black with white markings. Lisa was thought to be about four years old and

Bibbi was older when they arrived at the shelter. At first, the volunteers could not pet or pick up either of these stray females, but over time the cats became less timid and fearful. When we began observing the feeding routines, we noticed that they hung out in an empty litter pan placed in an area bounded by a wooden frame under the kitchen furnace and next to the dry food bin. As we stood by the doorway to watch the feeding, they attracted our attention by staring and stretching their necks toward us. Janet reached down and stroked Bibbi, whereupon Lisa strained toward her to be petted as well. We noticed that the empty litter pan was almost always provided for them, and the two never left the litter pan at the same time. If the cat guarding the litter pan was not in the pan, she was right next to it. On one occasion when Lisa was eating elsewhere in the kitchen and Bibbi was outside the pan, Garbo jumped into the pan. Bibbi gave her a hard stare and she quickly jumped out and slunk off as Bibbi stared after her. One day, according to our field notes,

> Upon arriving at the shelter we notice that Bibbi and Lisa are missing and that their empty litter pan is not in its usual place. One of the copresidents happens to be there and Janet tells her about their attachment to the empty litter pan. The president asks one of the volunteers to bring over a clean litter pan and as soon as it is in place, Bibbi and Lisa materialize and sit in it.

Carly, a well-liked tiny feral cat, often sat outside the litter pan on the periphery of Bibbi and Lisa's friendship. Bibbi and Lisa accepted her proximity, but we never saw her attempt to enter the litter pan. Thus, all of the evidence suggests that the litter pan served as a "home base" or "nest" that was protected and maintained by the two friends. They never brought food or any other items to it. Rather, they primarily rested and slept in it or watched it from just outside. It was important or *meaningful* to them. Thus, a major element of their friendship was the maintenance of the "nest" as a cooperative endeavor. We did not witness any other instances of this behavior during our observation of the shelter. In addition, Bibbi and Lisa rubbed each other, rested and slept together, and groomed each other. Bibbi was dominant over Lisa in regard to food, but Lisa was not otherwise subordinate to Bibbi or other cats:

> We bring some moist food for Bibbi and Lisa, who show a marked preference for it. Bibbi eats. Lisa turns around and tries to eat, but Bibbi pokes her and stares at her hard. Lisa turns around and sits. Bibbi leaves the food and sits on the dry food bin outside their "nest." Lisa turns around, washes, and eats.

By the end of 1997, Bibbi's health began to deteriorate, leading to a variety of separations and reunions with Lisa as Bibbi was periodically hospitalized or placed in a cage in the infirmary to be medicated and recover:

> Bibbi is out of isolation. She and Lisa seem thrilled to be together again. Bibbi is licking Lisa (instead of swatting her) and Lisa is trilling and rubbing against Bibbi. I pet both of them, and normally retiring Lisa begins to swat me. Is it because I stopped petting, is it competition for affection from me, is she protecting Bibbi and keeping Bibbi to herself? Steve goes over to pet Bibbi and Lisa races over. Is she saying, "Bibbi is mine"? I'm leaning toward that interpretation. I've never seen Lisa so assertive.

During December 1997 the board closed off the space under the furnace that had been Bibbi and Lisa's home base. They did this because cats were using that opening to defecate, urinate, or vomit behind the row of appliances, which was almost impossible to get access to for cleaning. As a consequence, Bibbi and Lisa became "homeless." Bibbi staked out a shelf for herself on a floor-to-ceiling cat pole between the front and middle rooms, but Lisa became disoriented. She hid under the refrigerator or the desk in the dining room, venturing out to eat from time to time. When she encountered Bibbi in the kitchen once, she showed no special recognition of her though we were reasonably sure that she saw her. We asked the staff about the situation, but no one we spoke to seemed aware of the problem. In February 1998 Bibbi and Lisa became ill at the same time and were caged together. During this period, their old friendship rekindled and they seemed very happy. Then they were separated again in March because stool samples were needed from each of them. In May the two were adopted together, but it was too late for Bibbi. She died at the end of June, leaving Lisa alone once again.

The litter box "nest" was a central feature of the friendship between Bibbi and Lisa. It is not uncommon in situations in which food is plentiful for feral females to form colonies in which the young are reared cooperatively in common nests.[19] These nests serve practical goals such as making it possible for mothers to obtain food while other mothers care for their kittens. Bibbi and Lisa's nest, on the other hand, served a social rather than a practical purpose. It seemed to act as a symbol of their friendship, the site in which the interaction rituals they shared took place. Without the nest, the shared reality of their relationship floundered. When they became ill and were caged together, the cage may have served as a surrogate nest that allowed the friendship to reassert itself briefly at the shelter. Unfortunately, we do not know how their relationship fared once they were adopted.

Bibbi and Lisa's friendship lasted from some time prior to the start of our observations in the summer of 1996 until they were adopted in May 1998 and, possibly, until Bibbi's death the next month. Another long-lasting friendship between a pair of cats was that between Merlin and Corey. Merlin was a large, black, feral male who could not be handled by humans, and Corey was a moderately sized black and white male who was skittish and shy but not truly feral. He could be petted if approached slowly and sometimes sought out humans for attention. Merlin was a convivial, even-tempered fellow who could not be described as either dominant or timid with other cats. Corey, on the other hand, was very low key and passive. These two began resting and sleeping together in whatever basket or cat bed was available in the fall of 1998 and were still doing so when the study ended in early 2000. They licked and washed each other as well, and on one occasion Corey followed Merlin out to the kitchen and drank from the water bowl Merlin had just drunk from. Theirs was a close relationship, but it never became as exclusive as that of Bibbi and Lisa's. Merlin still occasionally slept with other cats as did Corey, and sometimes other cats joined them. But they did sleep together more than they slept with others, and they also demonstrated a high regard for each other.[20]

There were many other friendship pairs, trios, and larger groupings at the shelter and, like that of Merlin and Corey, these friendships were mostly nonexclusive and overlapping. We have numerous instances in our field notes of Lucky, an orange feral male, licking, head-butting, and snuggling with both Lotus and Philip. Philip, in turn, was often seen sharing a sleeping spot with Tasia, who herself had numerous other regular sleeping partners, including Corey. Among the free cats and particularly the old-timers, then, there was a network of shared friendships of varying intensity and duration.

Many interrelated factors affected the longevity of relationships at the shelter. A major one was the adoptions, which removed the friendliest cats from the shelter, sometimes even before they had the time to form close friendships. When they did form such friendships before adoption, they were usually adopted out together as previously indicated. In addition to the shifting population as a consequence of adoptions, friendships were also broken up by illness, death, and actions of the staff, which sometimes unintentionally interfered with the social life of the cats. One more example in addition to the closing off of Bibbi and Lisa's nesting space will suffice. One day when we entered the shelter we saw Ariel, Danny, and Marney (all feral/semi-feral males) sleeping together in a small bed for one cat in the living room. When the feeding began, they woke up and went out to the kitchen where they began eating dry food together from an attached two-bowl set, sides touching and rubbing as they ate. On two other occasions these same cats slept together in the much-coveted Garfield bed, also suitable for one cat. Then one day someone removed the beds from the living room. So, Ariel joined the group who slept on the feral shelf in the litter room, Danny took up residence on the kitchen food cabinet, and Marney returned to his practice of sleeping with as many different cats as possible.

The longest friendships tended to occur among feral cats who were at the shelter for long periods of time, often until their deaths. Also of interest, because it goes against popular belief, is that in our observation the strongest friendships were among cats of the same gender. Males and females did rest, sleep, and wash each other, but we did not observe them to develop persistent ties, with two possible exceptions. First, Brianna, a small, feral, dilute calico, often associated with Lil'Guy, a small, feral, gray and white male, until we adopted him. Then, for a while, she and Logan, who was almost identical to Lil'Guy, were always asleep together at the top of a cat pole in the living room when we entered the shelter. After several months, the relationship seemed to wane, and they no longer slept together. Second, as will be discussed at some length in Chapter 6, Sonnet, a three- to four-year-old female, befriended Virgo, a woebegone juvenile male, and their friendship persisted until Sonnet was adopted. Finally, it was difficult for exclusive ties between pairs of cats to develop at the shelter because other cats often wanted to join in. With 50 to 60 free-roaming cats in three small rooms and a litter room, it would have been a full-time job to keep all interested parties away all of the time. Only a few cats, such as Bibbi and Lisa, seemed to have the stamina or interest to do so.

The picture that emerged was one of a multitude of overlapping friendship pairs and friendship groups. It is our contention that this network of interwoven friendships was a central element in the social organization of the cat community in the shelter. This was in sharp contrast to what is found in feral cat colonies, where the social organization tends to be based on kinship. More specifically, it is based on relationships among and between "matrilines," which consist of a female and one or more generations of her direct descendants.[21]

The Culture of the Cats

We conclude, first, that a distinctive culture emerged out of the interaction among the cats at the shelter–a culture that fostered affection, friendship, and social cohesion. The norms of this culture appeared to be binding on the majority of cats at the shelter to the extent that personal space had become a minor consideration for most of them. These norms were transmitted to the new cats, most of whom quickly shed their initial fear and hostility. We identified three major ways that shelter norms were passed on to new cats. The first was the human-initiated system of caging new cats in the main public rooms where the free-roaming cats hung out. This insured maximum contact between caged and free cats, who would scamper up and down the cages and rest on top of them. The second was the cat-initiated interaction that occurred between caged and free cats. As noted in Chapter 3, the free cats often responded to the distress of caged cats by coming over and being a calming presence for them. Relationships formed between caged and free cats who would regularly visit with their caged friends. Sometimes caged and free cats would play through the bars. The third mechanism was the sanctions, most often positive, that supported the peaceful norms of the shelter. Any cat who accepted the norms of the cage tops, for instance, might expect to be groomed, have someone to snuggle with, or find a safe, comfortable environment in which to take a nap. Occasionally negative sanctions were imposed on misbehavior as when Marquis stopped Tucker from swatting Beatrice as noted below.

We also concluded that the distinctive culture of the shelter was related to three conditions of shelter life. The first was the absence of competition for food, the second was the safe environment of the shelter, and the third was the crowded conditions of the shelter. Looking first at food and safety, let us consider Abraham H. Maslow's hierarchy of needs for humans.[22] He argues that humans first seek to satisfy their need for food; then they seek safety. Only after these primary needs are met will humans then look for belonging and love. Like humans, cats have a broad range of needs, which have different priorities under different conditions. Also like humans, for cats food and safety take precedence over other needs. The shelter, remember, admitted primarily stray and abandoned cats. To them, the plentiful, regular, and varied food supply contrasted sharply with their previous lives. In addition, the shelter was a safe environment, and cats were protected from the major sources of disease and injury they faced as strays, including injury from sexual competition. Because their practical needs, and the goals they gave rise to, were satisfied,[23] the cats sought to satisfy social needs and goals, just as in the case of humans. In particular, the cats looked for physical contact, affection, friendship, and pleasure. Even feral cats attempted to pursue such social goals when possible through the formation of colonies[24] and through regular social gatherings such as those observed by Leyhausen in Paris.[25]

The shelter staff often observed that the cats who adapted least well to shelter life were those who had come directly from homes because their human companion had died or they had experienced another misfortune. For these cats, all of whose needs were being met in their previous homes, the shelter was not an occasion for the emergence of "higher" needs or tolerance for physical closeness as it was for the stray cats. During our period of observation, a cat named Bandit came into the shelter from precisely such a situation.

While he was caged, he hissed and snarled at other cats just as Susie did later on. Although he was let out of his cage when he stopped doing that, at first he sat at the kitchen door swatting every cat who passed by. Other times he sat by himself looking lost. He did slowly begin to adapt, but we were all thrilled when he was adopted because everyone recognized his unhappiness. His new human companion indicated that Bandit was quite content.

The third factor involved in the dominance of social needs and goals among the shelter cats was the crowded conditions at the shelter. Except for brief periods when the rate of adoptions was high, the shelter was always at capacity, which could mean approximately 50 to 60 cats freely moving about three small rooms plus the litter room and another 10 to 20 cats in cages. In the absence of competition for food and in the safe environment of the shelter, the cats seemed to use the crowding as an opportunity for close physical contact. This is at odds with the findings of Leyhausen, who argued that cats placed in crowded conditions would become aggressive.[26] Leyhausen's data, however, came from experimental situations that had the effect of maximizing anxiety and insecurity in the studied population. The shelter cats, on the other hand, were often rescued from threatening situations, nurtured by shelter staff until they were recovered, gradually socialized to the shelter before becoming part of the free population, and made as comfortable as possible. Some of them had prior experience as members of a feral or household cat colony.

Thus, the shelter was characterized by a high degree of social cohesion rather than aggression. This cohesion was fostered by the cats' freedom to pursue social goals, the emergence of norms of tolerance for physical closeness in a crowded setting, and an extensive network of overlapping friendship groups. This cohesion was not limited to cats who were kin, as Tabor[27] found, since only a small number of the shelter cats were related. It was also not restricted to juvenile cats, since most of the cats we have been discussing in this section were mature adults from approximately four to ten years old. And, finally, it was not restricted to cats who knew one another as juveniles, since almost all of these cats came into the shelter as adults and met at the shelter. It was, we argue, the environment of the shelter that facilitated the display of these feline capacities to a greater degree than other previously studied settings.

Finally, we conclude that the cage tops, beds, and other soft spots in the shelter represented to the cats not just comfortable places for sleeping, but also safe places to relax, to find intimacy with others, and to explore their need for affection and friendship. This shared meaning is one that emerged through interaction among the shelter cats themselves. Thus, the cats redefined objects in their environment, which can be viewed as a form of symbolic interaction. In addition, Bibbi and Lisa's litter box nest, which served no practical purpose, was symbolic of the shared reality of their friendship. This friendship had difficulty finding expression when their nesting spot was closed off.

Social Roles at the Shelter

So far we have focused primarily on the norms that emerged out of the interaction among the cats at the shelter. In this section we examine the social roles that also emerged in the cat community. We define a role as a specialized activity performed on behalf of the group. We have identified three such roles–two of them were initiated by the cats while the third one required human intervention.

The first and more prominent of the self-initiated roles was that of the "mascot" or "guardian" of the shelter, which was played by Marquis, a middle-aged, black and white, male cat. Marquis was a stray who had been fed outdoors for three years by one of the shelter officers. When he developed an abscess on his leg she brought him into the shelter. He was a very affectionate cat at the shelter, often climbing into an available lap or asking to be picked up and held. Because of his history, however, the volunteers (and we) were a little wary of him. He had been adopted out three times and returned each time by people who claimed that he had attacked them without provocation. One of his adopters brought him back following a trip to the emergency room to have slashes on her arm treated. She came into the shelter with her arm bandaged up to the elbow. A second adopter brought him back, claiming that he had "pushed her down the stairs." Although it is possible that these stories became exaggerated over time, the shelter officers deemed him ineligible for adoption in the future. They interpreted these "attacks" as Marquis indicating that he wanted to come home to and was happiest at the shelter.

The volunteers recognized Marquis's "mascot/guardian" role, and introduced visitors to him as the shelter "guardian" or "security guard." Marquis was a very prominent presence at the shelter. He always seemed to be in the thick of things, greeting people and overseeing all activities. Because Marquis was the official mascot, his photo or image often appeared on items made and sold by the shelter, such as note cards and calendars. Once the shelter staff arranged to have Marquis execute several paintings, which they sold at

Security Guard Marquis keeps a watchful eye on the shelter.

a local bookstore in conjunction with the store's promotion of the book *Why Cats Paint.*[28]

We highlight two specific aspects of Marquis's "guardian" role. First, he almost always "checked out" new cats, or returned cats. We have noted several incidents when Marquis either jumped on top of the cat carrier or he sniffed and peered into the carrier as if to "inspect" new arrivals.

We see the second aspect of the "mascot/guardian" role in Marquis's protective behavior toward other cats in the shelter. Generally, whenever a cat cried, Marquis investigated and, sometimes, took action. One of the cleaner/feeders described an incident involving Marquis and Glorianna–an elderly cat who preferred to stay in a cage rather than mingle with the general population. On this occasion, Glorianna cried in protest when the cleaner/feeder took her out of the cage to clean it. She said that Marquis ran over and "attacked" her leg because he thought she was hurting Glorianna.

We also observed this behavior directly:

> I [Steve] went over to change cage tops alone. Kate and Doug were there. I noticed that Sunshine was no longer in the cage with Cedric [where she had been placed for socialization] and was on top of the big cage. Kate said that she had escaped when one of the cleaner/feeders was doing the cage. When I climbed the ladder to check the blanket on the big cage Sunshine jumped down on the small cage and Kate saw a chance to recapture her. She moved in and got Sunshine by the scruff, but Sunshine was hanging on to the cage wires with her claws. I was able to free her and Kate put her back in the cage with Cedric. While all this was going on, two women came into the front room from outside and in the confusion, one of them stepped on Hal's tail. He screamed and headed into the middle room. Marquis, who was in the front room, went charging after Hal and seemed to pounce on him although he did not appear to be "attacking" him but, rather, making sure he was okay. I moved into the middle room to also check on Hal when Marquis suddenly charged me and clawed at my pant-legs. I interpreted this as an effort to "protect" Hal from further harm. Kate came in and picked Marquis up and sat him on her lap in the front room. He calmed down right away.

We observed another example of this behavior, this time involving three cats who were rescued from a neighborhood house that had burned down. The cats were quite dirty and needed serious grooming. When Kate and Betty tried to clip the nails of one of them, the cat cried in protest. Immediately

Marquis appeared and pushed his way in to see what was happening. We observed one further incident of this type in which Marquis appeared to play his "security guard" role by sanctioning a cat for violating the norms against aggressive behavior:

> I'm told that Tucker has attacked Opie. Both of them were just recently freed from the living room cages. Well, Tucker comes running into the kitchen and swats another cat. I think it was Beatrice. Marquis attacks Tucker and drives him off.

The second self-initiated role that we identified was that of the "companion," as played by Marney. Marney was a male tiger who, like Marquis, had been at the shelter throughout our period of observation. He was one of the feral cats and, therefore, not readily adoptable. Although the shelter staff did not recognize Marney's "companion" role as such, they frequently commented that Marney was "friends" with many of the cats. We discovered Marney's role by systematically observing the association patterns among the cats. We have already described the tendency of the cats to snuggle together on the cage tops and in the beds and baskets and we have described various "friendships" that formed at the shelter, including one involving Marney. We found in our field notes that Marney slept with, snuggled with, or spent time with several cats. It was as though he was a friend to all, and whoever needed a companion could turn to Marney. The following are typical observations for Marney:

> We arrive at the shelter and none of the cleaner/feeders are here yet. The cats are mostly sleepy and not too active. Marney and Laura are being very affec-

> tionate with each other. They are rubbing against each other and butting heads. A little while later Chelsea is in a bed designed for one cat. Marney climbs in with her and slowly forms himself around her. He goes to sleep while she continues to doze without interruption.
>
> Marney gets in one of the boxes with Corey. A little later Tess climbs in and nuzzles Marney and licks Corey. Marney responds and begins to lick both Tess and Corey.

Although we focused on Marney here, two other cats also played this "companion" role. The shelter staff mentioned that Lucky, an orange feral male, was an object of affection from several cats, which we also observed. Chelsea also performed the "companion" role mainly by allowing a variety of cats to snuggle with her in the beds and baskets in the shelter.

The performance of the third role, the "socializer," required human assistance. Cedric, a black feral male cat, was initially caged when he arrived at the shelter. Two of the shelter staff developed a fondness for Cedric and worked with him to get him used to people. When he was ready to be let out into the general population, he showed himself to be quite timid and resisted leaving his cage. Since cages were generally in short supply at the shelter, the only way they could "indulge" him was to put him in with other cats. When they did this, they discovered that he had a calming effect on his cagemates: New cats and feral cats would snuggle next to Cedric and allow themselves to be touched and petted by humans. The staff people began to call him Uncle Ceddie, and referred to him regularly as "our socializer."

Sunshine, an orange female, generally stayed in the front rooms and would not allow anyone to touch her. Here is what happened when she was placed with Cedric:

> We come into the shelter and, as usual, check the cages for new arrivals. We observe that Sunshine, who is normally loose, has been placed with Cedric. Betty explains that Sunshine is in with Cedric to be socialized. She opens the cage door, reaches in, and begins petting Sunshine. She says that Sunshine would never have allowed this a week ago.

Cedric, who was eventually adopted by a shelter volunteer, was particularly good at socializing "wild" kittens:

> We are observing in the front room when Harriet, one of the shelter officers, comes in and calls our attention to Cedric. They have put him in a cage with two new feral kittens–Paris and Pippen–in order to socialize them. Harriet says that within a week they will be able to handle them.

Many animals routinely divide tasks among themselves when it confers a survival advantage. Male vervet monkeys defend the troop by chasing off intruders. And older female giraffes watch calves while their mothers forage for food.[29] In the shelter we saw individual cats take on roles voluntarily because they found them congenial. Victoria L. Voith and Peter L. Borchelt also mention animals, including cats, who sometimes take on roles in the group on their own volition.[30] Swanson provides the example of Simon, who served as a mouser on a British frigate shortly after World War II and was wounded in a communist attack on his ship on the Yangtze River in China. In spite of his wounds, and long before recovery, Simon resumed his mouser duties and regularly visited the wounded sailors in the ship's hospital. He had numerous opportunities to abandon ship, but never did. For his service, he received a ribbon from the captain and a medal from the British when the conflict ended.[31]

We argue that the development of roles in the cat colony at the shelter parallels role development in informal human groups such as friendship groups. Central to such roles is some specialized activity performed on behalf of the group that initially emerges out of members' personalities. For example, someone who is a natural leader begins to assume leadership responsibilities, and someone who is naturally nurturing begins to engage in mothering behavior. Institutionalization of the roles advances when they take on a normative dimension, that is, when the group develops a collective expectation that certain members will perform these specialized activities. And institutionalization is complete when sanctions, either positive or negative, are invoked to insure performance of the role. Both Marquis and Cedric performed specialized activities for the group consistent with their personalities, but it was the volunteers, rather than the cats, who both expected and positively sanctioned their role behavior. In Marney's companion role, his significant others were the cats. They expected him to be companionable, and they rewarded him by reciprocating his friendliness and by seeking out his company.

Dominance and Aggression?

Our observations during feeding time confirmed our main finding: In this shelter cat colony, norms of affection, friendship, and social cohesion dominated. It was, in fact, touching to see such a large number of cats in one small room peacefully sharing and enjoying their food. The little aggression that occurred involved very few cats who tended to be new cats whose aggressiveness diminished as they became fully socialized to the shelter norms.

Feeding and Eating Routines

Feeding took place in the kitchen twice a day. At each feeding the free-roaming cats were given wet cat food, the quantity of which varied with the size of the population. Fresh dry food and water was also put down at the same time. The wet food was placed in six or seven dishes from eight to ten inches in diameter, and usually these were arranged in the middle of the kitchen floor so that cats could gather around them. More than one can of wet food of different flavors might be placed on each plate. The volunteers spread out the bowls of dry food and water around the edges of the room.

Because the food appeared at regular intervals, the cats knew when they were to be fed. When they saw a volunteer at the expected time (but not otherwise), they anticipated the feeding by heading for the kitchen and milling about in the room. There might be anywhere from 15 to 30 cats in the 12′ by 12′ kitchen, so it could be quite crowded. During this *prefeeding phase,* the cats tended to be extremely social, butting heads, touching noses, rubbing against one another, and locking tails. Their facial expressions appeared happy, hopeful, and alert. Those who had a special eating place took their position at this time. There were occasional perfunctory swats and a few raised backs among the cats, but, again, the swatting was done by either newly integrated residents who were still uncomfortable surrounded by so many cats and still uncertain of the norms or a small number of

Ferals and friendlies share a meal peacefully in the kitchen.

cats who were "troublemakers" or "prima donnas." During the period of observation the most consistent troublemakers were Sid Vicious, Mandy, Dotty, and Penny Royal. Sid was not so much feral as cantankerous, Mandy and Dot were both feral females, and Penny Royal was great with people but not with other cats and she was later adopted. The prima donnas were Cimmaria and Kemet. Cimmaria, the shelter's oldest cat, died at 20 years old, and Kemet was adopted.

During the *feeding phase* most of the cats shared their plates contentedly, and even the troublemakers sometimes ate peacefully with other cats. Their aggression tended to be very perfunctory and seemed as likely to involve their friends as anyone else. There was also a lot of moving about from plate to plate, presumably because the different plates contained different flavors:

> Alex [volunteer] comes in and many cats move to the kitchen. Cimmaria is on the counter and Kemet is on the sink. . . . Danny rubs against Alex. Mandy slaps Lucky [good friend of hers] when he passes. First plate down–five cats eat–Penny Royal, Lucky, Danny, Garbo, and a black cat I can't identify. . . . Philip wanders around, sniffs butt, stretches. Second plate down–Bibbi, Mandy, and a tiger eat. . . . Very calm scene. Lucky and Philip butt heads in friendly fashion. Duncan comes to Mandy's plate. Her ears go back. He moves to a plate with Dot and Druscilla. Philip eats now with Penny Royal and Duncan. Mandy and Penny Royal stare at each other. Penny Royal goes off. Mandy hits a tiger who wants to eat.

Both the volunteers and the cats acted to minimize feeding time aggression. For instance, the volunteers all had been "trained" by Cimmaria and Kemet to feed them separately in a special place. This made them happy and peaceful. Also, when a cat wanted to eat from an already occupied plate and thought there might be resistance, that cat would usually creep up to the plate very slowly and delicately eat from the side opposite the occupant. Or he or she might pull some food off the plate with a paw and eat it on the floor. This behavior generally satisfied the occupant.

Some cats, such as Cimmaria, Kemet, and Mandy had particular "eating styles." Cimmaria demanded to be fed on top of the counter that the volunteers used to open the cans of food. She tasted each offering, selected her favorite, and then hunkered down to eat. Most volunteers obliged her. Kemet only wanted to eat in the sink. If he did not get his way, he cried, followed the volunteer around, swatted other cats, and made a general nuisance of himself. It did not take him long to "train" the staff to feed him at the sink. Mandy would come into the kitchen and give a shrill, single-note cry and stare at whatever cat might be at a plate from which she wanted to eat. She was almost always successful in getting a plate to herself in this manner. There did not seem to be much genuine emotion or hostility behind Mandy's actions. She was not excited when she gave her cry and stared. Hers was more feigned emotion or "emotional performance." Mandy had friends, but she wanted a plate to herself if she could get it.

We do not believe that the cats we have labeled "troublemakers" and "prima donnas" were dominant cats exercising their dominance over others when they tried to get a plate for themselves. First, these attempts were irregular. Mandy would give the shrill cry to get her own plate one minute and eat with her friends the next. Second, these cats did not try to dominate in other situations in shelter life. Third, attempts to hold on to a plate were fleeting. Some cats refused to give up their plates, or the troublemaker might change plates to sample other food. Finally, we never saw a pariah cat in the shelter. In the laboratory situations characterized by Leyhausen[32] as producing dominant cats, there was always a cat who was subject to much more harassment than the others, called the pariah.

As feeding time drew to a close at the shelter, many of the cats engaged in *posteating rituals,* which involved washing either individually, with a friend, or washing that friend. These baths might take place in a sleeping spot, and after the bath the cats might take a nap or go off to play:

> Some cats are still eating, but the "feeding frenzy" has died down. Duncan washes himself as does Danny on the periphery of the plates. Mandy, intent on washing her back, rights herself just in time to prevent a fall off the washing machine. Several cats have moved out of the kitchen into the middle room where they sit washing themselves. Merlin and Corey wash each other's necks in one of the cat beds while Marney, Logan, and Alice all wash themselves and each other more or less simultaneously in another small bed as it teeters on an unsteady chair. At the same time, Marquis races around the room chasing a ball.

Although most cats ate peacefully, usually sharing plates with no sign of tension, more aggression was associated with the *prefeeding* and *feeding phases* than there was with resting either on the cage tops or in the beds. One cause might be found in fluctuations that occurred in the distribution of food when the shelter was short staffed or volunteers were distracted. For instance, we observed a couple of incidents in which a staff member repeatedly raised the expectations of the cats that they were going to be fed and, then, became distracted and did not feed them. This resulted in stress and increased aggression by the cats toward both the humans and one another. Another cause might be found in the history of the cats. Remember, most of these cats came from disadvantaged backgrounds in which they were malnourished or even starving when they were rescued. We believe that even though the cats knew that the food would come on a regular basis at the shelter, there might be the residue of a survival mechanism to defend food. Thus, some cats might initially respond by defending their food by swatting or some other sign. Then they would realize there was no need to do that and go back to eating peacefully. This was consistent with the fact that after a swatting incident, the cats often went on eating together. It was also consistent with the ritual of creeping up to an occupied plate to avoid startling the occupant. Finally, as we mentioned, the new cats or "troublemakers" did most of the swatting.

Socially Challenged Cats

Since the interactions among the cats at the shelter were overwhelmingly characterized by sociability, intimacy, egalitarianism, and a remarkable disregard for personal space, cats who were loners, aggressive, sought to be dominant, or tried to defend a territory would be deviants whose behavior required some explanation.

Almost all of the deviant cats were newcomers who had just been released from the socializing cages in the living and dining rooms. Within one to four weeks, their conduct became indistinguishable from the majority. Many examples could be given of this phenomenon:

> Lewis's cage is open for him to come out. He makes only one try and goes right back in. Megan lets Tiggy and Katie out. Tiggy explores. Katie is unsure and stays near the cage. These are cats who have been in the living room cages, and Megan feels they are ready to join the community.... Tiggy is not doing well out of his cage. He is abusive to other cats. I suggest he be recaged and let out gradually when people are here. They prepare a cage to return him to.

> The food has been put down–one cat to a plate. Nothing interesting, except Tiggy who is free again and picking on everyone.

> They tell me there was only one adoption yesterday at the Pet Clinic–it was Camille. But Lewis was the big surprise. He sat for four hours on a table never moving except in the direction of his food dish. Tonight, however, he gets into a brief shouting match with another male. . . . Tiggy has calmed down considerably from earlier this month when he was first released into the free population.

Within one week to a month, all of these cats were socialized to the norms of the shelter and were no longer involved in conflicts with other cats. All of them have since been adopted. Thus, the aggression noted here was *situational* and subsided as the new cats became a part of shelter life. The changing behavior of new cats over time also highlighted the fact that the norms of the shelter were a result of interaction and, in time, molded new individuals in their image.

A handful of cats at the shelter seemed to lack social skills and either adapted more slowly or incompletely. Sid was one such cat. Someone had trapped him and brought him into the shelter in 1997 because the person was concerned about his welfare. He was a medium-sized, male tiger who always looked slightly electrified because his fur tended to stand up on end. He was the crankiest cat we observed at the shelter, always getting into conflicts with others over sharing food, space on the cage tops, or just personal space. He played no favorites, swatting anyone, cat or human, who offended him. He did not, however, generally seek out situations of conflict nor did he do more than swat the offender. If he was eating and another cat or human came over, he would swat them. If he was on the cage tops and another cat or human woke him, he would swat the offender, though recall that he did briefly rouse himself to pursue La Salle on the cage tops. But this was the only time we saw such a response. Most interestingly, he never sought to separate himself from other cats, always sleeping and eating among large groups. He was the shelter curmudgeon, and we may call his behavior a *reactive* aggression. In 1999, his temperament began to improve somewhat, and occasionally he allowed himself to be petted without swatting the volunteer. He also became less irascible with the other cats, allowing them to cuddle with him on the cage tops without incident. Eventually, he was adopted but was returned because of landlord disapproval.

Emilio and Smokey were far more actively aggressive than Sid. Despite their aggression toward other cats, both cats were tame and got along well with humans. Emilio was a very large, all white, long-haired male stray who was probably on the street a long time judging from his condition on arrival at the shelter. He generally rested and slept alone, usually near a window. Once he was released into the general population, he had to be placed in the "time-out"[33] cage repeatedly and was only allowed out when volunteers could watch him. He started fights with other males and was obsessed with sweet-tempered Nathan, making his life at the shelter miserable. He was adopted on three occasions. The first adopter returned him because he beat up her other cats and the second because he did not like her home and kept trying to run away. The third adoption seems to have been successful, so far.

Smokey, a beautiful, sleek, all dark gray male, was a "return." That is, he had been adopted as a kitten, kept a year and a half or so, and, then was returned to the shelter when the owner or some member of her family developed allergies to him. Like all such cats, he expressed a great deal of unhappiness at the shelter. He was very combative and had to be placed in the "time-out" cage repeatedly from June until December 1999. He actively went after many other cats, including Marquis, who, as we discussed, played a significant role in shelter life. Like Emilio, Smokey picked out a cat for special and continuous harassment, a loveable female called Ruby, who he reduced to living on a cabinet ledge to avoid him. As soon as he was freed from a cage, he would search for her, look at her fiercely, and try to attack her. We would take him from the room, try to play with him, try to give him food, reprimand him verbally, and push him away, all to no avail. He was like a boomerang, returning to harassing Ruby the second he could get back to her until we caged him again. This behavior stopped only when Ruby was adopted in November 1999. By January 2000 he finally seemed to be calming down, and we saw him several times peacefully sleeping with other cats on the cage tops or eating with others quietly.

Both Smokey and Emilio seemed to be *true aggressors* who sought out conflict and wanted to dominate more passive cats. There were few cats like this at the shelter, and we could not always find a cat with these characteristics, even though the shelter drew a large portion of its population from stray and feral cats. In fact, all of the truly aggressive cats we can recall over the years were tame cats in their relationship to humans. We have no way of knowing if these cats' personalities were molded by experience or nature. Smokey did seem to change, which suggests the former. He had been an only cat for the first year and a half of his life except for whatever time he had with his mother and siblings. He may simply have lacked the skills necessary to relate to other cats and had been too unhappy to learn them.

Thus, we have identified three types of aggression at the shelter. One was strictly situational, associated with new cats whose aggression disappeared as they became acclimated and took on the norms associated with life at the shelter. A second type of aggression is reactive, in that it was a response to the behavior of other cats who did something to offend the aggressor. A third was a more active aggression in which the aggressor sought out conflict with other cats. We call this true aggression, because the cat did not respond or responded only partially to the benign norms of the shelter.

Although all of the aggressors we have described were male, and males were more heavily represented among aggressors, females could be aggressive as well. There were a number of females who would attack males as well as other females and who presented themselves in an aggressive fashion. Panda, for instance, was a particularly aggressive cat:

> Panda is out of her cage. She is not happy and hisses and goes after the other cats. She won't come into the kitchen [which is crowded with cats] to eat with the others. I give her a separate plate at the door away from the others and she eats. . . . Panda is back to hissing and running after the other cats. She growls at Marquis near the cages in the living room. She is very friendly toward people.

Panda's aggressiveness did not subside in the usual period of time. After several months, she was adopted but then returned. The couple who returned Panda seemed unable to live with her "a minute longer." Because she never acclimated to shelter life in the first place, having her return after an absence did not help the situation, and she continued to show aggression to other cats. Finally, she was again adopted and was not returned during the time of the study.

As with the males, most female aggressiveness was situational and restricted to newly freed cats. Panda might have been a case of true aggressiveness, but, in our experience at the shelter, we have never had to place a female in the "time-out" cage or keep such a cage available for a female. We never had a female who picked out another cat for special harassment as Smokey and Emilio did. We know, however, that this phenomenon exists, because it occurred in our own home. Our cat Shelly hated our oldest cat, Annie, from the moment she set eyes on her and attacked her ferociously at every opportunity. None of our considerable efforts to alter the situation worked, and we can only say that as time passed she became less enthusiastic and regular in her harassment of Annie, who was a timid and mild-mannered cat (Annie has since passed away). Shelly's background was much like Smokey's. She was the only cat of a woman who left her at the shelter after three years when she moved to Florida. We adopted her before she joined the free population and, thus, when Shelly came to live with us she had not interacted with other cats since kittenhood. She was distraught over her situation and did not acclimate to us or the other cats for nearly a year. After a year she did become friendly with all of us except Annie. Thus, at least some instances of true aggressiveness and dominance have their roots in the social background of the cat who is socially inept as a result of a lack of experience with other cats. Further, the unhappiness of such cats at losing their home and human companion(s) reduces their motivation to adapt. Why some of these cats attempt to create a pariah of one of the cats in their environment, we do not know. Shelly was successful in doing this in that she influenced our other cats to treat Annie in the same way. The efforts of Emilio and Smokey were not successful, we think, because with so many cats at the shelter it was very difficult for any one cat to command such influence. In addition, Emilio and Smokey's activities were often disrupted by the volunteers who caged them if they were harassing other cats too much.

Finally, researchers have noted that cats lack fixed dominance displays. When cats do exhibit dominance it tends to be restricted to a specific activity or place such as Bibbi dominating Lisa over food or a cat dominating a pathway at a specific time of day.[34] Next, we discuss territoriality, which is closely linked to dominance.

Territoriality in Shelter Life

Territoriality arises out of the need of animals for a secure food and water supply and may be defined as the area an animal defends. This is in contrast to a home range, which is the larger area in which the animal lives.[35] Shortly after birth and before their eyes open, kittens demonstrate a bond with their nest by finding their way back there from a short distance away.[36] Thus, location or territory may be fundamental to a cat's nature. Territory, in turn, is linked to dominance because, by definition, territory is defended and maintained through aggression. Territoriality, then, would appear to be an anticohesive force that would work against friendship and social intimacy.

All of the evidence we have, however, indicates that territoriality in domestic cats is negotiated, a product of interaction, and does not exclude the possibility of social ties. For example, Leyhausen argues as follows:

> Important with regard to territorial behavior is the fact that domestic cats are less repulsive to one another than their wild relatives and in most cases can be brought to share a home area and often even the first-order home with one or more other cats. At first this might seem to be a serious disadvantage, but probably is it simply that the special circumstances mentioned above have brought out more clearly the cohesive factors within the population which are certainly at work in wild populations as well. As stated above, it is quite normal for the pathway network of neighboring cats to overlap, and overlap in this case means the common use of pathways, hunting grounds, and sometimes other amenities such as sites for sunbathing and lookout posts.[37]

Use of common pathways and amenities may occur at different times of day.[38] If the cats meet, however, or a cat is using another cat's resting place or lookout post, they will not necessarily fight. Even if a cat has shown his or her superiority at another time, he or she will not necessarily drive the inferior cat away. Further, cat territory markings do not function as deterrents to other cats who will sniff markings in a leisurely fashion and are not intimidated by them. Thus, at best, the territoriality of domestic cats is weak in that it does not override other considerations, and dominance is highly relative and limited by time and place and activity.[39]

We begin our examination of territoriality among the cats at the shelter by looking at how they viewed the outside environment. We never observed the shelter cats become excited when they looked out the windows and saw other cats outside. Stray cats abounded in the shelter neighborhood and the volunteers fed them on the house steps. As for new cats inside of the shelter and in the living room cages, the residents were very friendly toward them. The following incident from our field notes illustrates the point:

> Megan tells me she has agreed to take in a 10-year-old cat whose owner has died. A woman relative brings Tigger in–a large, silver tabby, very handsome. She tried to keep him, but she has two dogs, one of whom wants to eat Tigger. She cries giving him up. Megan puts him in a cage in the living room since he is a house cat who has all his shots. He immediately faces the back wall and begins to howl in distress. The woman is distraught and cries harder. We try to comfort her. Meanwhile, Pumpkin and Smedley go to the cage very quietly and sit down in front of it. They watch Tigger intensely. He realizes he is being observed and turns around. Pumpkin and Smedley are very calm and quiet in their demeanor. Well, Tigger calms right down, stopping his howling. Soon he begins to explore his cage.

The shelter cats often related to the caged cats in a variety of ways, comforting them, playing with them through the bars, and, yes, stealing their food if the plates were placed too close to the edge of the cage. Thus, the shelter as a whole seemed not to be defended as a territory by the shelter cats.

Inside the shelter, the closest thing to a defended space we witnessed was Bibbi and Lisa's litter box nest. The only other defined space associated with a group of cats was the "feral shelf," which was a very high shelf near the ceiling of the litter room. Cats accessed it by climbing some lower shelves. Many feral cats used this shelf, and a few remained there when humans were at the shelter. Others stayed there when they were newly released but later made their way out to other resting places and divided their time among a variety of spots. Although this was definitely a preferred location for feral cats, and most of the cats on this shelf were feral, there were usually visitors from the tame population. These cats were always welcomed, and some were regular guests. We never saw this shelf defended by the ferals. They definitely saw it as a place protected from the human members of the community, because they knew the humans could not reach the shelf easily. We also did not see the various beds and sleeping spots around the shelter defended. Some cats may have had a preference for a particular bed–the Garfield bed was very popular–but it was never defended. Many cats liked to change sleeping spots from time to time. As previously noted, we did observe occasional hissing on the cage tops between old-timers and newcomers, but nothing came of it, and no one was driven off.

All told, we can only say that there was some tendency to bond with a location such as the feral shelf or a particular sleeping spot, like a kitten with its nest. But such bonding was not a bid for territory and did not result in a defense of the location. We separate the concepts of bonding with a location from territoriality because they are not the same thing and do not serve the same purpose. In a population of cats in need of a food and water supply, it may not be possible to distinguish these phenomena. In this population of well-fed cats, however, some cats persistently wanted a special location, particularly the feral cats, but they did not meet the definition of being territorial, because they did not defend the space. The only possible exception we encountered in our four years at the shelter was the case of Bibbi and Lisa with their litter box nest. We will continue this discussion in the chapter on the feral cats.

In conclusion, the anticohesive force of territoriality is relatively weak in the domestic cat to begin with in that a cat's territory is very much subject to negotiation and the give and take of interaction with others. At the shelter this force was so weak that the more cohesive dimensions of feline personality and social organization came to the fore and dominated. As we have previously mentioned, the variables of shelter life that weakened territoriality were the constancy of the supply of food and drink, the safety of the environment, and the generally crowded conditions.

One of the implications of this discussion is that although domestic cats may have an innate program, or instinct, to hone in on a location they associate with safety, they do not have an innate program to carve out and defend a territory. Whether cats define a territory is totally variable depending on their environment. Under some conditions, such as scarcity of food and water, they may be able to draw upon their innate location program to create a territory. But they do not have to do this, and they can put this ability aside completely when it is not called for, as in the setting of the shelter. Alternatively, they may negotiate the terms on which they hold the territory, as in the observations of Leyhausen[40] and Tabor.[41] As Barber notes,

> Most instinctual programs can be shaped and sharpened significantly by feedback, practice, and learning to fit closely the demands of particular circumstances. The flexible implementation that is needed to fit constantly changing conditions requires choice (judging or deciding between alternatives), which in turn requires intelligence.[42]

The Everyday Life of the Shelter Cat

There is little doubt that strong elements of social organization existed in the shelter cat community. The first element was a social structure that was overwhelmingly egalitarian and rooted in a network of overlapping friendship groups. The effective absence of hierarchy and dominance that we have described essentially allowed every cat the freedom to fully participate in shelter life. The only notable social division that we found was that between old-timers and newcomers, and such a division was by its nature temporary, like the division between old and young in human communities. A lack of hierarchy did not mean a lack of differentiation insofar as social roles were also part of the social structure of the cat community. Some of these were community-wide, such as Marquis's role of "shelter guardian" and Marney's role of "companion." Others existed between and among particular cats, such as "friends" in friendship pairs or groups. These roles were not performed out of any necessity, but, rather, developed out of the cats' individual personalities. The second element of social organization was a distinctive shelter cat culture with norms that guided everyday life. These norms included expectations that food, litter boxes, and sleeping spots were to be shared and aggression and hostility were to be discouraged. Add to these the strong norm of tolerance for physical closeness, and the result was a cooperative culture in which the cats' "higher needs" for affection and friendship could be realized.

Just as they do for us, the cultural choices made by the shelter cats simplified their everyday decisions. They knew that they could move about the shelter freely and settle in any available comfortable spot. If they sought physical closeness with other cats, they could reasonably expect a friendly reception. When meals were served, the cats knew that they did not have to wait for a dominant cat to finish eating first, and they knew they did not have to fight to protect a dish for themselves. They could simply all join in and share the plate. Most cats shared this culture at the shelter, and they transmitted it to the new cats as they were socialized into the community. The cat community also had its symbols in the investment of the cage tops and the various sleeping spots with a shared meaning for the whole group. Anyone who climbed to the cage tops or entered a basket or bed could find affection, intimacy, and safety there. In all, the shelter cat community exhibited a high degree of social cohesion. Aggression, dominance, and territory, the three variables emphasized in animal behaviorist studies, played an insignificant role in the life of this shelter. Cats, then, are highly flexible and adaptive animals with a great capacity for a rich social life.

The Cat Community and the Social Self

In Chapter 3, we examined the ways in which the cleaner/feeders fostered self-awareness and development in the shelter cats using, amongst other things, Dawkins's five criteria of such awareness as a guide:[43] (1) the ability to adapt to novel circumstances or to find new solutions to new problems, which suggests the ability to learn; (2) the ability to learn from others as well as from one's own experience; (3) the ability to make choices in situations that provide more than one option; (4) the ability to cooperate; and (5) an overall complexity of behavior. We must now ask if the cat community we have described in this chapter also contributes to the enhancement of the social self in the shelter cats.

A perfect example of Dawkins's emphasis on the ability to adapt to novel circumstances as evidence of self-awareness is that the shelter cats resorted to intimacy and social cohesion as a solution to the overcrowding at the shelter. Time after time each new Whiskers cat had to learn to put aside notions of aggression and territory and live peacefully and affectionately in a large colony of other cats. Most of them did just that, changing themselves in the process.

This responsiveness to others certainly included the ability to learn from others, Dawkins's second noted indicator of self-awareness, as when feral cats learned to trust the shelter through their association with Cedric, one of the socializer cats. We witnessed additional examples, such as when new caged cats learned to trust other cats from the attentions free-roaming shelter cats gave them while they were still caged. And newly freed cats learned that, as long as they behaved properly, they would be welcome on the cage tops and other sleeping spots. This ability to learn from others, however, has another dimension that Dawkins did not emphasize. The underlying factor in all such learning from others is the more inclusive sociological notion of "taking the role of the other," or seeing things from the perspective of the other. The ability of the cat to form close friendships with other cats *who are not kin* presupposes the ability to take the role of the other and to learn from the other. In other words, since the friends were not biologically related to one another the basis of their affinity was most likely social. The friendship represented a cat's choice to have special ties to another cat, which could only be achieved through taking the role of the other (taking account of one's friend in one's own behavior). The ability to learn from others is also central to culture, which Dawkins discusses.[44] As we have noted, at its heart, culture is all that is shared by a group. Inherent in such sharing is the transmission of knowledge among members. In addition, friendship, in its own right, contributes to self-awareness by helping to locate the individual in the community. It is a small structure within the whole that helps place felines just as hierarchy places canines in canine groups.

The free-roaming shelter cats also made many other choices. Cimmeria chose to have special foods given to her in a special place, and Kemet and Danny also chose to eat in a special place. As we will see in Chapter 6, cats often chose their adopters by insistently relating to them when they came to the shelter. When the shelter cats responded to overcrowding with affection and social cohesion as opposed to aggression or isolation, this may be viewed as a choice as well. The same is true of their eating peacefully together from communal plates, instead of choosing to wait and eat in shifts so that everyone has his or her own plate. We agree with Dawkins that the choices animals make are a window to their inner lives, helping us to understand their needs, desires, and mental capacities. Had the shelter cats made a different set of choices, a shelter such as Whiskers would not be possible.

There were a number of instances of cooperative behavior at the shelter, another of Dawkins's criteria of self-awareness in animals. Bibbi and Lisa's efforts to maintain a "home base" was a major instance of cooperative activity between the shelter cats. Other activities that may be viewed as "cooperative" would be when free cats chose to "help," befriend, or comfort caged cats, which was a common occurrence. Although we have already discussed this under the criterion of learning from others, this behavior has multiple implications. An additional implication is that the free cats who were helping were making a choice. Marquis's behavior also had multiple implications. When he "helped" fellow residents in distress he was making a choice as well as cooperating with the cat community to maintain order.

Last is the Dawkins criterion of complexity of behavior. There can be little doubt that the shelter cats' behavior was complex. The emergence of norms, sanctions, friendships, and social roles, and the development of shared meanings or culture in the cat community all suggest a rich and varied social life. We argue, then, that the quality of the social life of the shelter cat was affected not only by the humans who were a part of the total community but also by the cats themselves. The cat community was itself a major source of the social self of the shelter cat and enhanced that self as much as did the human shelter volunteers. Before leaving the cat community, however, in the next chapter we take a closer look at the feral cats and their special contribution to the society of the shelter.

5

The Feral Cats and Shelter Solidarity

IN PREVIOUS CHAPTERS we referred to feral cats at the shelter. In this chapter we examine these cats in more detail to see what they can teach us about cat behavior and the shelter community. The term *feral cat* is generally applied to domestic cats who are born and raised independently of humans. Those who have studied or observed such cats over an extended period[1] generally find that, contrary to the stereotype of cats as solitary, most feral cats live in colonies. The size of the colonies ranges from six to sixteen members,[2] and they usually congregate near a food source such as a restaurant, dumpster, or apartment building where residents regularly feed the strays. Thus, although feral cats live independently of humans, they are heavily dependent upon human food sources. Our best resource for information on the composition of feral cat colonies comes from a national survey of cat feeders and cat rescuers carried out by the Massachusetts chapter of the Society for the Prevention of Cruelty to Animals (M/SPCA) and the magazine *Animal People.* They estimate that roughly one-third of the cats in feral colonies are abandoned pets, and the remaining two-thirds are the "true ferals," cats born and raised completely away from humans.[3]

At the time of our study, approximately 30 to 40 percent of the cats at the shelter were considered feral or semiferal. These were cats who were living independently of humans at the time they were admitted to the shelter, and who remained unsocialized or only partly so. That is, they did not allow themselves to be handled, and they generally resisted being touched or petted. Some of them, like Lucky and Sunshine, simply retreated if someone approached too closely or extended a hand to pet them. Others, like Tasia, hissed and bared their teeth and occasionally struck out with their claws (as one of the authors discovered when he shed some blood testing the limits of Tasia's "personal space"). There is no way of knowing whether these cats were "true ferals" or strays who became feral. We do know, however, as we describe below, that many of the cats who came into the shelter as ferals became socialized either partially (the semiferals) or completely through their contact with the volunteers and, possibly, with the friendly cats.

Shelter policy on feral cats evolved over the years. The shelter officers at first routinely took in ferals, but since these cats could not be adopted out (except in special circumstances) they threatened the ability of the shelter to take in new cats. Indeed, the ferals who remained unsocialized were clearly the long-term residents of the shelter. Feral cats also presented problems if they became ill and needed to be medicated on a daily basis. Carly, for example, would not allow herself to be medicated and would not eat any food that had medicine added to it. She eventually had to be euthanized. And Misti, a tiny tiger feral who hid most of the time behind the refrigerator, became ill and died there before she could be tended. The shelter eventually tried to place limits on the number of ferals it took in, but it could not exclude them altogether. They sometimes came in accidentally when a group of strays was rescued or when homeless kittens were taken in with their feral mothers. Sometimes volunteers who fed stray cats in their neighborhoods pressured the shelter to take in a feral cat who became ill or injured. In the almost four years of our observations in the shelter, we saw only a few new feral cats admitted to the shelter. Feral kittens were a different matter, of course, since they could almost always be socialized and adopted out.

Ferals and Friendlies in the Cat Community

Two bulletin boards at the shelter contained pictures of all the current residents. One was for the "ferals," and the other was for the "friendlies." We will use the term "friendly" to describe shelter cats who were oriented to, and could be handled by, humans. The pictures were important because staffers needed to know who was approachable and who was not. The cats themselves, however, made no such distinctions in their relationships with one another. The line between friendlies and ferals was regularly crossed as the resident cats formed friendships and gathered in snuggling and sleeping groups. Consider the following description of the cage tops from our field notes:

> We enter the shelter and it is cold in the front room. Zoe [friendly] is on the big cage with Scamper and Philip [ferals]. On the smaller cage are Nathan and Amberlee [friendlies] cuddled together with Logan, Tasia, Laura, and Sunshine [ferals].

We mostly observed both ferals and friendlies together on the cage tops. This mixing of feral and friendly cats was also seen in the beds, baskets, and sleeping spots:

> In the dining room, Chelsea [feral] and O'Malley [friendly] sleep together in the Garfield bed. Next to them, on a cushion, Marney [feral] and Lil'Momma [friendly] sleep together. Then, Lil'Momma actually gets on top of Marney crossways and rests her head on his back. Merlin [feral] comes over and looks to get into one of these scenes, but there is no room for his large presence. He is very put out about it and keeps looking, hoping for a possible spot he can fit into.

Although the cats themselves did not distinguish between friendly and feral in their everyday interactions, we contend that the ferals made up the solid core of the cat community–a community we describe as emphasizing friendship, affection, and physical closeness. We base our claim on the previous chapter and our discussion of socially inept cats. These few cats were loners, aggressive, dominating, or territorial–cats who deviated from the norms of sociability, intimacy, and egalitarianism. We indicated that these cats were almost all newcomers who had not yet been socialized to the norms of the cat community. They shared one other attribute in common: Virtually all the socially inept cats were friendlies. As a group, then, the feral cats appeared to be more likely than the friendly cats to integrate into, and accept the norms of, the cat community.

There is additional evidence that the feral cats exhibited a greater degree of sociability within the cat community. In our observations of the beds, baskets, and sleeping spots, for example, feral cats were almost always present, either with other ferals or with friendlies. We rarely found friendlies cuddling with other friendlies in these places. Friendly cats were more heavily represented in the cage-top groupings, but even here we noticed that when these groupings became larger (three or more cats), ferals were far more likely than friendlies to be in the majority.

We noted earlier that homeless cats often live in colonies. This means that they would have sought out other cats for social interaction and affection[4] and would be familiar with and comfortable operating in a cat community. If our shelter ferals tended to come from colonies, they would have had a "head start" in adapting to the shelter cat community. With this idea in mind, we interviewed the shelter staff regarding the origins of the shelter ferals. Of the 46 ferals we became acquainted with during the period of observation, the majority (29) came from colonies, seven came from the homes of collectors or were found individually, and the origins of the remaining ten could not be determined with certainty. Most importantly, we learned that all of the most social cats, the ones who appeared in our notes again and again as forming friendships and snuggling with other cats, came from colonies. These were also the cats primarily associated with the enhancement of the social self in the shelter felines.

We argue, then, that feral cats, as prior members of a colony, came into the shelter having learned to trust other cats. They sought comfort in the other cats almost as soon as they were released into the free-roaming population, and they readily accepted, and later enforced, the norms that guided the shelter cat community. Remember, also, that the ferals were most likely to be the "old-timers" in the shelter and, thus, the most settled in the cat community. Unlike the ferals, the friendly cats were somewhat less wedded to the cat community because they could and did find friendship and affection with the human members of the shelter community. This was confirmed, as each time we entered the shelter the cats with whom we formed relationships came running to be petted, picked up, and played with.

Socialization of the Feral Cats

Ferals and humans had an uneasy relationship with one another. From the perspective of most ferals, the shelter must have seemed a very good place. It was warm, food was plentiful, and they quickly learned the norms that would enable them to integrate into the cat community. The only serious stumbling block was the humans. Safety from humans was achieved in a variety of ways. The most important of these was the previously mentioned feral shelf, which many of the ferals occupied while humans were present. Some old-timer ferals, such as Lucky and Peridot, were almost always there at such times, but it was also a preferred spot for the newer ferals such as Pinky who were very frightened of humans. Cleaning Pinky's cage took courage and was the subject of many anxiety-generated jokes. When she was released into the general population of cats, she quickly found the feral shelf, where she stayed for weeks hissing at anyone who even looked in her direction. Pinky was a beautiful, long-haired, dilute calico. Trudy, one of the volunteers, was very attracted to her and decided to work with her with the intention of taking her home. We interviewed Trudy about her decision to adopt Pinky:

> I came on Fridays when I didn't have shift duties. Pinky was always on the feral shelf. I climbed a lad-

Lucky and friends find safety on the feral shelf.

> der and talked to her. At first when I put my hand out near Pinky, she would hiss but she would always stay and not jump down. The other cats on the shelf would scatter as soon as they saw me, but Pinky never left. After a couple of months, I began stroking Pinky's face. She would back away but stay. I never felt afraid of her. That is where things stood when I decided to take Pinky home.

During this period when Trudy was relating to Pinky at the shelter we all noticed the difference. Pinky no longer hissed at us when we went into the room with the feral shelf, and she began venturing out into the other rooms, staying on the cage tops while the volunteers cleaned around her.

Many of the ferals graduated to the cage tops and other high places in the shelter after they got over their initial fear of the humans. Eventually, some of them would even eat with the humans around as long as they did not come too close. In general, the flight sensitivity of the ferals was much greater than that of the friendlies. The ferals ran first and asked questions later. The friendlies were much more likely to wait to see what was going to happen before they decided whether they would run.

The shelter volunteers expressed strong concern for the well-being of the feral cats. They worried about them, and many volunteers attempted to work with particular ferals, although it was very difficult to find the time during shift work. Some volunteers felt they could do the most good by adopting a feral. They came to the shelter on their own time to socialize "their" feral. Ferals were a challenge to many of the volunteers who felt they had a special affinity for cats. They related tales about their success in touching or otherwise engaging ferals. Philip was a special challenge for several volunteers. He was a dark gray cat with bright green eyes who came from the home of a collector. As mentioned, collectors take in many more cats than they can care for. Often animal control officials are called in to remove neglected and sick animals from collectors' homes. When Philip arrived at the shelter, he wildly hissed and spit at volunteers, permitting no human contact. He was extremely sociable with other cats, however, and formed close ties with the other ferals. In time, he also became attached to the friendlies of the cat community. He was a regular on the feral shelf and, eventually, became at home in the whole shelter. Volunteers saw Philip as an irresistible challenge, often calling him by name and attempting to touch him. He responded vigorously to his name. When anyone said, "Philip," he would stop and turn in the direction of the speaker, staring and looking very alert. Anna, a Wednesday evening volunteer, often attempted to touch him:

> Anna is here. She pets Philip in the kitchen near his butt. He looks uncomfortable but doesn't hiss. I say that he always hisses at me if I even go near him. She says you have to approach him from behind.

When Philip was ill and had to be caged and handled, almost anyone could touch him:

> All the volunteers are commenting that Philip is letting everyone pet him. He is recently back from the vet and is caged in the front room by himself. Betty calls us over to show us he can be petted. We try it successfully. Nora does the same and others follow. Poor Philip, normally you can't get near him.

Among shelter volunteers, Kate was the shelter "specialist" in handling ferals. Her sensitivity to and affinity for the cats was truly remarkable. She could handle ferals no one else would even think of approaching. Although most of us did some work with the ferals, Kate's efforts were legendary. Through the socialization period, she kept them confined in a cage and dependent on her and the other volunteers for comfort and support. She took ferals who were all claws and teeth, snarling and spitting, and held them in her lap without gloves, petting them and talking to them until they became calm.

She used their own signals to reveal her intentions. For instance, cats often squint their eyes at you when they feel affection for you. Kate found that they responded in kind when she squinted her eyes at them once they were calm enough to pay attention to her signals. She was responsible for socializing Cedric, "the socializer," who carried on her work with feral cats. Her most remarkable achievement, however, may have been the case of Tara. Betty, one of the copresidents, told the story:

> Tara was the nastiest cat on the face of the earth. When she was about 10 years old [she had been at the shelter for years], she became ill and was so sick we were able to cage her and treat her–she couldn't resist. As she recovered she was very dirty from the medications and not grooming herself during her illness. So, one night I come into the shelter and Kate says, "Lets try to give Tara a bath." I thought she was crazy, but she insisted and, so, we bathed her and cleaned her ears–something every cat hates. Well, she began purring and turned into a "big mush ball," loving the entire experience and she's been a big mush ever since. So much so that Melanie and her husband adopted her.

Almost as remarkable were the cases of Michael, Monte, and Mallory, three feral juvenile cats taken in by a volunteer. She put them in her basement, but she was unable to handle or make any progress with them. Kate said she would try, and told her own story:

> When I got there, Mike was in the fuse box, Monte was in the ceiling, and Mallory was in an old, discarded stove. They were about seven months old. We captured them, somehow, and when I got home I put them in my foster room. You've seen it. There's no

> place to hide. I kept them for four months before bringing them down to the shelter. I made a pest of myself. I held them and made them sit in my lap every day. I talked to them all the time I was with them. I also made a point of eating with them when I fed them. You know, I think what goes through a feral cat's mind–they think we hunt too and maybe eat cats. I brushed them, trimmed their nails, played with my fingers around their mouths. I continued to work with them at the shelter. It took close to a year.

At the time of our observations, one would never have known they had been ferals. Michael greeted us at the shelter door and took food from our hands. He always wanted attention. Monte climbed our legs so we would pick him up and smooch with him. Mallory, the female sibling, was adopted.

We also have a success story to tell. When we met Lil'Guy at the shelter, we could hardly touch him. He shrank back as far as possible and closed his eyes tightly, perhaps envisioning himself on our table as a tasty morsel! We worked with him for a year, making steady progress, although there were setbacks if we were absent for a short time. After about three months, he stopped shrinking from our touch, and by about five months he was greeting us at the door when we arrived. At this point, he began paying special attention to us, seeking out our company. Soon he began indicating that he did not like our attentions to other shelter cats and, as we described earlier, would come between us and other cats when we were petting them. Lil'Guy also loved to play with us with a string, but the one thing we could not do was pick him up:

> Lil'Guy has a drippy eye. Megan asks us to try to get him so she can medicate him. I try to pick him up,

> but he's immediately alert and wary and will not let me. He is right behind Steve. I ask Steve to try it. He tries and Li1'Guy panics and scratches his head and neck and runs off. Steve is upset though not seriously hurt. I'm upset because I realize how feral he still is and that we can't adopt him if we can't pick him up.

Not all efforts at taming were successful. Philip never yielded to the efforts of the volunteers beyond accepting a brief pat on the rump, and even Kate could not tame Lucky and his sister Lacy although they were only five months old when they arrived at the shelter. One problem was the lack of free cage space. Kate said that if you cannot confine them, you cannot tame them. Several things that the volunteers did seemed to have the effect of enhancing the relationship between humans and feral cats who were past the more pliable kitten stage. Clearly, confining them and becoming the central figure in their lives who feeds them, cleans up after them, and comforts them was very important. Handling them daily, even against their will, also seemed to be a crucial factor. Kate touched them extensively all over their bodies when she held them, including their faces and paws. Notice that illness often could be the occasion for such handling with older adults such as Tara or Philip. Talking to them was important in bonding, as it established common words with shared meanings. Feral cats often did not like to hear humans speak initially. It frightened them. Eating and playing together were also bonding tools. We found playing very effective in that few cats could resist a string even if they were frightened. No one thing was critical in all cases. For instance, although we never confined Lil'Guy in order to relate to him, we made substantial progress with him. In fact, he became so friendly toward us that we decided to adopt him even though we still could not pick him up. To take him home we caught him in a dead sleep and placed him in the carrier before he knew what had happened.

Thus far we have discussed what humans did to establish a relationship with ferals. Ferals, however, might also initiate friendly interactions, often by choosing a location where they would allow the interaction to take place. The volunteers who adopted cats who remained feral said that their adoptee did this after they settled in at home. Trudy adopted Pinky and Mandy while they were still quite feral. Speaking first of Pinky and then of Mandy, Trudy said:

> Under the dining room table: I can lie on the floor next to her there and rub her face. She won't allow such contact when she is under the couch. That's her other hangout.
>
> She must be under something or someplace where she can get away.... The special place is the spare bedroom.... She gets on top of the cabinet and hangs out and squints her eyes at me. I can pet her there. Also, under the desk....

When we asked Libby if there was a special place where she and her feral, Laura, interacted, she said:

> Oh yeah. In front of her private heat vent and her private chair or looking out the back door. It's much easier to pet her in those places than elsewhere. Anywhere else, she thinks I'm going to pick her up and bolts.

Our own feral cat, Annie, who was adopted at four months, ran and jumped up on our bed or the radiator cover in our bedroom when she wanted to express her affection for us. In those places we could hug and smooch with her at will. If we tried to even pet her elsewhere, she would bolt. Only when she was past the age of 10 or so did she include other locations. Even at 17, she would bolt from us if the spot was not right.

We believe that this behavior relates to the safety of the nest described in Chapter 4. There we noted that many feral cats could only risk the danger of relating to a human from the safety of a certain location that acted as a nest substitute. This would further suggest that the tendency to gravitate to a nest has less to do with territoriality than it does with safety. Thus, the cabinet top, the bed, and the other spots described above allowed these ferals to explore novel relationships with humans from a safe haven. The symbolic nature of these gestures may be inferred from the lack of objective safety many of these locations provided. The case of Bibbi and Lisa described in Chapter 4 is a bit more complex. Their litter box nest also allowed them to reach out to humans and explore the possibility of human friendship, but the nest was primarily an expression of their friendship for each other. This may account for their unique efforts to defend it as territory and keep it for themselves, which was not found in any of the other cases we have described. Not every feral cat, however, selects a special location for human interaction. The one exception in our experience was Lil'Guy. He did not have a special location for interacting with us at the shelter and, when we adopted him, he did not choose one at home. He allows us to pet and play with him wherever we happen to come together.

We must add one final note on the socialization of ferals. We believe that the friendly cats also contributed to this process in several ways. For example, if the feral cats sought to associate with friendlies, they were drawn into the central rooms of the shelter. This put the ferals in proximity to humans who, as we have described, often talked to them and attempted to relate to them. It is also likely that the ferals observed the friendlies in interaction with humans, which may have allayed some of their anxiety. We discovered that if a friendly and a feral were snuggled together on a cage top, we could often use the friendly cat as "cover" and reach over him or her to pet the feral.

The Special Roles of Ferals, Friendlies, and Humans in the Shelter Community

In this chapter, we have described a number of differences between the ferals and the friendlies in the shelter. In the process of doing so, we discovered that each of these groups had its own special role to play in the shelter community. The ferals' role was to maintain the norms and character of the cat community as we have described it. They were the glue holding the cat community together. This role was grounded in several factors. First, the ferals were the long-term residents of the shelter and, as such, they had developed a greater commitment to the cat community. Second, the ferals, by definition, were not oriented toward humans, so that in seeking friendship and affection, they would turn to their fellow cats. Finally, for many of the ferals, this tendency to seek community with other cats had been reinforced by their prior experience of living in colonies. Given this situation, we believe that the current shelter policy of not admitting ferals, if carried out, will eventually change the character of the cat community in the undesirable direction of greater conflict.

The friendlies' role was to provide a link between the cat community and the humans. They drew the humans into the shelter community by seeking affection and comfort from them and by forming relationships with them. (It is no wonder that volunteers were a major source of adoptions from the shelter!) They also, we believe, drew the ferals into the larger community by helping to socialize them. The shelter volunteers, of course, played a major role in the socialization of the ferals and in maintaining the human orientation of the friendlies. They spent many hours forming ties to the shelter cats by cuddling, playing, and talking with them in addition to the time they spent tending to their basic needs.

Our focus on the feral cats has reinforced a major theme of this study, which is that all the members of the shelter community were active "makers" of its culture and social structure. This enterprise was not simply a human operation. It was a product of the interactions among humans, friendly cats, and feral cats. Given the importance of each of these constituencies, the loss of members from any group represented a potential disruption to the total community. We now turn to the issue of member loss and how the Whiskers volunteers coped with it.

6

Leaving the Shelter Community

THE SHELTER COMMUNITY was a community in flux as both human volunteers and cats came and went. We talked about the "comings" earlier in our discussions of the ways in which new volunteers and new cats were integrated into the community. In this chapter we focus on the "goings" and the potential they had for disruption of the shelter community. By "goings" we mean volunteer attrition, cat adoptions, and cat deaths.

When Volunteers Leave

Volunteer attrition was relatively high, but this did not disrupt the human–cat community because those leaving the community tended to be the newest members. Those who decided that they no longer wished to volunteer seemed to realize that quickly, often disappearing after the first few weeks of service. The leadership of the organization, however, remained largely constant except for the resignation of the original president in 1991. One of the copresidents who replaced her remained in that position since 1991, and the other resigned in 2000. Her replacement was a long-time volunteer. Many of the other original volunteers, such as the health coordinator, the major socialization agent, and the fund-raising staff, continued to perform those duties throughout the period of our observations.

The continuity of the community endured also because replacement volunteers were not very difficult to find. There were many people in the immediate area who shared Whiskers's philosophy regarding animals and few organizations that operated on those principles. Thus, many volunteers were grateful to work with such an organization. They would have found it unthinkable to work in a regular shelter in which healthy animals were euthanized for lack of space.

Finally, many volunteers might leave a particular activity, such as cleaning/feeding, but they remained with the organization in some other capacity or continued to contribute to Whiskers financially or to follow its fortunes. Volunteers often felt committed to the cats and the goals of the organization even when their life circumstances no longer permitted any regular activity such as cleaning and feeding. One couple, who were cleaner/feeders for years, retired from their work positions and took to the road. Yet they still contacted Whiskers when they were in town and came to help out if needed. For many, volunteering for Whiskers was a life-altering experience.

We regret that we have no direct data regarding whether the departure of particular humans affected the shelter cats. It would have been impossible to separate out this one factor from others in the shelter that affected cat mood and condition. What we do know is that particular cats were very happy to see particular humans when they came to the shelter, even after an absence, and indicated regret when particular humans left the shelter after a shift. In Chapter 3 we noted that Marquis greeted us at the door after a three-week absence and wanted to be picked up and petted as before we left. This was not conclusive, as Marquis was friendly toward many people. Lil'Guy and Amberlee, however, never greeted anyone but us to our knowledge and sometimes looked dejected when we left:

> As I come in Amberlee comes to greet me. I pet her. Lil'Guy comes in to the room, hears our voices, and looks up. We pet him. Later Lil'Guy, Roxanne, Bogie, and Angelina play with me with the string. No one is aggressive even when they bump into each other. Roxanne plays wildly. Lil'Guy loves the game too. When I stop playing, I pet him and then leave. He looks dejected.

Cats certainly have the memory and emotional capacities to "miss someone." We have witnessed cat reunions with one another as well as with us after we return home from vacation.[1] As indicated, however, we have only the above type of evidence on the impact of volunteer attrition on the cats. As for the cat community itself, we saw nothing that indicated any impact from volunteers leaving. The departure of any particular volunteer would not have affected the organizational policies or cultural practices that supported the cat community. Thus, we believe that the cat community remained unchanged as volunteers came and went.

Saying Goodbye to the Cats

Between 1998 and 2000, Whiskers found homes for 802 cats, or an average of 267 cats per year. Whiskers set a record of

334 cats adopted during 1997. That year they had a retail location for adoption clinics that was very successful, but the company changed its policy on permitting such clinics on its premises. Since many of the cats that Whiskers took in were not immediately suitable for adoption because of illness, injury, or undersocialization, the commitment to them represented a major investment of time and money. Medical bills ran into the hundreds or even thousands of dollars as veterinarians repaired broken bones from road accidents, mended severe wounds inflicted by humans or other animals, and treated disease caused by malnutrition and neglect and other problems too lengthy to catalogue. For instance, a worker rescued Hot Sticks (a.k.a. Electra) from electrocution at a power substation. She was at the veterinary clinic for nine weeks receiving skin grafts and other treatments associated with burns. This care ran into the thousands of dollars. She recovered with only the loss of her left ear, which had been burned off. Thus, in addition to Whiskers's philosophy, which made every cat valuable, every cat was indeed precious by virtue of the investment of time and money that he or she represented. Veterinary bills were about $70,000 per year or approximately three quarters of Whiskers's annual budget.

Whiskers's adoptions took place in the shelter itself or at adoption clinics. These were two very different venues in which the dynamics of adoption were played out. At the shelter cats could be active participants in the adoption process, and their presentation of self to the prospective adopter was a major factor in successful adoptions. At the clinics, which took place in a variety of retail settings in the region, the cats were caged and had to play a more passive role. In addition, most of the kitten adoptions took place at clinics and, when kittens were available, the adult cats attracted less attention. Another way in which shelter and clinic adoptions differed was that when prospective adopters called Whiskers, they were screened on the telephone before being invited to the shelter. Then, at the shelter, they were questioned further before a contract was concluded. At the clinics, there was no opportunity for the first screening. Everything had to be done on site, often with more than one prospective adopter being screened at a time.

Although we attended clinic adoptions, we focus here on adoptions at the shelter in which the interaction process was critical. The shelter was the most common site of adoptions until the last year of our study when the clinics were given greater emphasis as a way of increasing the speed and volume of adoptions. When prospective adopters came to the shelter, they were confronted with 50 or 60 cats from which to choose.

We found that people chose a particular cat for five reasons. The most common of these was the belief, often supported by the staff, that the cat had chosen them:

> One potential adopter is looking for a low shedder because of allergies among her friends. She wants advice on who to adopt. She has no other cats. She was initially attracted to Violet, but Violet did not "connect" with her. She finally chooses Petula because Petula "wanted to go home with me." She seems very happy with her choice. Then one of the volunteers comes into the room and says that cats know who loves them; that one of the shelter cats she took home waits for her and will only come out at the sound of her voice. Betty is still tending potential

adopters. Among them are a father and son who decide to adopt JJ and Kateria. JJ walked into their carrier after they thought they liked him. That clinched it. Kateria hung out on the father's neck while he was walking around and he liked the way she looked as well (a tortoise shell with white).

A prospective adopter comes in. I miss the beginning of the exchange because I'm watching Lil'Guy and Corey. Apparently, he came earlier and chose Mystic. Betty wasn't here then, but Megan told her, "Mystic was all over him." He couldn't take her then because he was going away for the weekend. He came back for her today and she went right up to him. He said he chose her because she was the friendliest of the cats to him. He was a young male of college age. She's a great cat. I hope he takes good care of her. Another adopter is here with his roommate. He's also a young male and he wants a playful, friendly cat. Megan shows him Barbara. He is mildly interested.... Then he looks at Black Amber in the cage and asks if she can come out. When she is out she rubs against him. He says he wants her. He calls his roommate over and says, "This is our cat. She loves us."

The cats varied in their responses to prospective adopters. Some ran whenever prospective adopters came to the shelter. They seemed to realize they were not shelter personnel. Others, like Marquis, greeted almost everyone and showed an interest in all shelter activity, including adoptions. Still others, like the cats in the above field notes, responded to prospective adopters selectively, taking an immediate liking to some while showing indifference or distaste for others.

Adopters tended to respond very favorably to the cats who showed an interest in them. This way they did not have to make a choice among the many cats available for adoption. The staff often supported their choice of cat, which made the adopters even more comfortable with their decision.

Sometimes it was necessary for the cat to choose all of the members of the household. One Sunday, a young woman came to the shelter seeking to adopt. Fairly quickly she bonded with Lester, but she did not want his sister with whom he was paired for adoption. In the end she could not make up her mind but returned the next Sunday with her own sister. She kept looking for Lester, found him, and picked him up. He was not that responsive and she clearly wished he were more so. Worse yet, Lester liked the sister even less. They finally choose Carla "because both of them liked her (they lived together) and she liked both of them."

A second important factor in adopter selection was the advice of one of the officers. Usually adopters would provide enough information about their situation so that the officers could suggest an appropriate cat for them:

> An adopter comes in with us. He has been at the clinic and talked about coming to the shelter to see a specific cat–Larkspurr. Kate talks to him about her. He pets her on the desk. He pets other cats too. He has another cat who does not like to be alone. He keeps trying to relate to Larkspurr who jumps to the floor. She is more interested in exploring. He is, in spite of the lukewarm reception, going to adopt her. Kate is filling out the forms. The adoption is complete with handouts and toys. He said he was following Kate's advice in taking Larkspurr. He just wants a cat

> for his cat. Kate wanted to be sure he knew Larkspurr's story–that she brought her feral kittens out for them to capture.
>
> The big news is that Nathan was adopted by the girl who was looking at him last Sunday. Betty said she took him on just her say-so because he would not relate to her at the time. Betty couldn't wait the usual week and called the night after the adoption to ask after him. The girl said that within two hours he was running around playing, eating, and making himself at home. It's hard to believe he was so depressed at the shelter. He just slept all the time and wouldn't respond when you petted him or anything. Remember that he had been adopted and was returned when his folks got divorced. The long stay at the shelter just depressed him terribly.

As in the case of the young man who adopted Larkspurr, people often wanted a cat as a companion for their cat at home. In this instance, compatibility was usually their major concern:

> A couple comes in to adopt. They want an older female to be a companion to their male. She can't be timid. We recommend Lillia and Rose. They pick up Rose and hold her. She is calm and stays near when they put her down. They said their male cat got along well with females.

A fourth factor in adoption choice was personality and temperament. This might be because of other cats at home, such as in the case above, or it might be for the sake of compatibility with the adopter and other human household members, such as in the case of the two sisters noted above who adopted Carla. Adopters might also be looking for a cat with the same characteristics of a loved cat who was no longer with them:

> When we arrive Betty is already there with prospective adopters–an older woman and a young girl. The older woman is the one who wants to adopt. She sees Willy and goes over to him and pets him. He responds by looking at her in a friendly way. She picks him up. Megan then shows her Marco. She finally chooses Marco and Tiberius. I ask why. She says Megan recommended them as like her 17-year-old male who died a month ago. He was laid back, gentle, and kind.

Finally, adopters sometimes chose a cat based on physical characteristics, such as age, color, and sex:

> An adoption has just taken place. A woman has adopted Bunny. I ask her why she chose Bunny (who was in a cage) and she says she wanted a young cat and she likes calicos. Now a mother and her teen-age daughter arrive. They are looking for a nice female according to Megan. She shows them Lucy and lets the mother hold her. Lucy is a big mush. Megan brings out Faith, who is shyer. The mother hands Lucy to her daughter and takes Faith. The daughter sits and holds Lucy. Jan asks why they are adopting, and the mother says she just lost her cat six months ago. They decide on Lucy.

Most of those who submitted to the interview and took the trouble to come to the shelter had made up their minds that they wanted to adopt a cat and were prepared to make a commitment. Occasionally, however, someone would pass the interview stage and appear at the shelter, but he or she clearly was not ready to commit to a particular cat:

> A prospective adopter arrives. She is a young woman who lives with a roommate. She says that she wants a calm cat who is one or two years old as her roommate will not tolerate a "crazy cat." Betty points out Ivy. The girl strokes her but wants to see others. Betty shows her Norris. She pets him and then a black cat. She goes into the litter room and tries to pet Felicia. Felicia shies back and Betty explains that those in this room are feral. She asks if she can just look around. She is looking at the living room cages now and asks to see one of the cats with the kittens in the middle cage–a calico. She wants to know her age. She wants to pick up the calico, who does not want to be picked up. Betty pets Mystic in the cage above. The girl then notices her and also pets Mystic, who is the friendliest of cats. The girl says, "You're a friendly one." She pets her some more and then closes the cage. She asks about the cats in the cabinets in the living room. Now she pets Tarot, who is also a very friendly black cat like Mystic. She asks her age, which is two years. Now she has gone on to another. She goes to Ophelia's cage and pets her. Ophelia bites her and slaps her. Steve and I can barely suppress our laughter. She is not put off but slowly withdraws. She goes back to the calico. I have taken a serious dislike to this prospective adopter and figure that she will not adopt. Finally, she says she likes tigers. We have only one–Bogie, who is not very friendly. When

we leave she is still there and I ask Betty if I can call her to find out what happens. [I call later and she did not adopt.]

Often prospective adopters came with a carrier or had made some other preparation to welcome their new feline companion into their home. In that frame of mind, they looked for signs that they were making the right choice. The best sign was one in which the cat made the choice for them by persistently interacting with them during their visit. Other factors could also help: the advice of a shelter officer who was experienced in the matter of matching a prospective adopter with a shelter cat, a sense of connection between a shelter cat and a cat of the adopters who had passed away, or the judgment that the personality of one of the shelter cats was a good fit with the adopter's home situation. If these factors did not provide the basis for a clear choice, the adopter could also use physical characteristics such as age, sex, and color to make a decision.

Adoption Ambivalence

Finding homes for Whiskers cats was the primary goal of the shelter, and many means were employed to promote adoptions. A newspaper ad ran in the major regional newspaper every day, and a local cable television station showed Whiskers cats who were ready to be adopted. In addition, a Saturday column in the feature section of the same regional newspaper provided information about adoption clinics, fundraisers, and particular animals associated with the various shelters in the area, including Whiskers. The media were very responsive when there were special needs cats who needed a home. When the leukemia positive Japanese cats referred to in Chapter 3 arrived at the airport, local television news shows covered the event and provided adoption information. Whiskers also ran three annual fundraisers, which garnered a good deal of public support. The clinics themselves were also a means of promoting adoption.

Shelter staff, however, felt some ambivalence[2] about finding cats homes. This ambivalence had several sources. First, no matter how carefully they screened prospective adopters, there was always some uncertainty associated with adoption. Staff always wondered, "Is this home really a good one?" Having rescued many of the cats from dire situations and having invested considerable time and money in restoring them to health and adoptability, Whiskers volunteers did not want to see them placed in danger or discomfort again:

> Megan is very frustrated. She has been speaking on the phone with an adopter who has seriously violated a provision in the adoption contract. Says she doesn't want to let any more cats out of the shelter because people are too stupid to be believed. They sign a detailed contract and it means nothing to them. Later a foster parent comes in with a yellow kitten who needs a shot. When Megan goes into the infirmary, the foster parent tells me that the shelter has a home for the kitten, but she is uncomfortable with it. The adopter has an old, territorial Siamese and a dog. The foster parent fears for the kitten. The adopter is coming at 7:00 and she wants to talk to her. I tell her to tell Megan about her doubts. She is reluctant even though shelter policy gives foster parents final say on adoptions.

> This is our first day back to the shelter after our vacation in Georgia. When we arrive, Nancy, Alice, and

> Betty are there. I ask Betty first thing about Bibbi and Lisa, who were adopted by a blind person. I was happy to find that she had reservations also about this adoption but says that, so far, they are okay. She was invited by the adopter to visit and will probably do so.[3]

In addition to the difficulty of judging the suitability of a home, even when the adopter had met all obvious criteria, volunteers formed attachments to particular shelter cats over time. Thus, adoptions sometimes meant losing a good friend. We provide two incidents in which we were involved, since our notes give the full flavor of the emotions that an adoption of a favorite can arouse in a volunteer. Amberlee, the subject of the first field note, was Jan's favorite and Eileen, the subject of the second field note, Steve's favorite:

> Megan has just arrived. No one else is here. She's waiting for an adopter. They come in–a young mother, her 2-year-old son, and the grandmother. They head for Amberlee–Please, don't let it be Amberlee. I suggest Smedley, who allows himself to be squeezed. Now, the kid is looking at poor Moses, who is 12 years old. The mother of the boy likes Pomeroy. She's holding him. The kid meanwhile is mauling Moses, who is unfazed till he kicks him. Then, Moses takes off. The mother tries to interest her son in Pomeroy. No dice. Now they are in the dining room. I stand in front of Amberlee in the hope they won't see her. The kid is running all over the place grabbing cats. The grandmother hauls him outside.... The mother chooses poor Pomeroy, who is young and funny and probably the best bet if you're

going to let 2-year-olds have a pet. She picked him because she likes the orange and he was real friendly.

An adoption is in progress with Megan–a husband, wife, and two dreadful boys, one about 6 years old, the other a toddler. They are looking at Eileen–Steve looks very upset. They are saying they would bring her back if she scratches. [What? The brats or the furniture?] Now the older kid is *in the cage* with the three sister cats rescued from the fire down the street. To her credit, the mother does reprimand him and haul him out. The toddler is screaming in the father's arms. They can't make up their minds. Whew! They leave. Thank God!

We were particularly concerned about adopting cats to households with children because of the children's erratic movements and sometimes-unpredictable behavior. As it happens, callbacks on Pomeroy indicated he was fine. The grandmother allayed our fears by telling us that the boy was in nursery school, so that he would not be with Pomeroy all the time. Eileen eventually was adopted by a nursing home with two other Whiskers cats. Each was assigned a floor at the home. As for Amberlee, we later adopted her for ourselves.

A final source of ambivalence arose because volunteers believed that each cat is a unique individual. Even though each adoption meant that a new cat could enter the shelter, no cat was ever replaceable. Volunteers conversed about past Whiskers cats and always welcomed news of them. When Marquis went home with a volunteer because he was becoming elderly, and the pressure of his security role was telling on his health, the volunteer provided weekly updates on his condition and responses. In addition, photos were displayed on the wall of past Whiskers cats who were adopted or died. As we show below, such efforts helped to minimize the potential disruptions to the human–cat community that adoptions might otherwise provoke.

Coping with Loss from Adoptions

Although adoptions were occasions of joy, they did represent a disruption of the human–cat community. The community was losing members, and their leaving often evoked some anxiety about their future well-being.[4] We identified a number of ways of coping to minimize the disruptions associated with cat adoptions. The first was the development of adoption procedures. The initial step in adoptions was the screening interview that we described earlier. The shelter officers designed this interview to distinguish potentially good caretakers from problematic ones. For example, they considered people who owned cats before and who kept these cats all their lives as good prospects for adoption.

Once the cat (or cats) had been selected, the adopter signed a contract with the shelter and paid a fee. In the contract, adopters promised to do such things as spay or neuter the cat, if this had not already been done, keep the cat indoors, and return the cat to the shelter if they were unable to care for him or her. Through the contract, Whiskers attempted to extend its control over the cat beyond the boundaries of the shelter. The fee, which was not insubstantial at $60, was designed to underscore the seriousness of the undertaking and to increase commitment to the adoption. The shelter officials were well aware, however, that these adoption procedures were often only partially effective as the following case reveals:

> During our visit to the shelter the phone rings and Megan takes the call. It is from a woman who adopted a kitten (Willow) from Whiskers last year and signed a contract indicating that the cat would be spayed at six months. She was calling, instead, to see if Whiskers would take the adopted cat's kitten! Megan was livid and said she would take back both cats. The owner was defiant and said she was taking the original cat to be spayed on Friday and would not give her up. Megan said she would prosecute her and hangs up. Just then Harriet comes in with two cats from the vet. Megan tells her about the call. Harriet suggests that they leave Willow with her, get verification of the spaying, and just take the kitten. It's left up in the air. I think it will be brought before the board.

The volunteers called the adopters to make sure everything was going well and offer advice if necessary, which served as the second coping tool. The staff called a week after the adoption and then again at one month, three months, and six months. Volunteers who had a favorite cat adopted were often anxious to hear how the cat was doing in his or her new home. In the following excerpt from our notes, we receive updates from one of the officers on three adopted cats:

> Megan says she has no information yet on Smedley, who was adopted at the clinic on Saturday by a colleague of Janet. Pomeroy [whose adoption by a family with a two-year-old boy was described earlier] is fine and loves the kid, apparently. Ivy, however, is in a home with birds and went after the birds. They wanted to give her back, but Megan told them to sep-

arate her when they can't watch them. She doesn't have the run of the house.

When a follow-up call revealed a problem, as in the case of Ivy, the staff made additional calls until the problem was resolved.

The adoption bulletin board was another way to minimize community disruption. Whenever a cat was adopted, her or his picture was moved to a special bulletin board for adoptions. This board was located near the door so that when volunteers arrived for their shift they could immediately learn who had been adopted. They did not have to wonder what had happened if a favorite cat was missing. This bulletin board was a reminder that finding permanent homes for the resident cats was a primary goal of the shelter and, thus, each adoption was an event to celebrate.

In an earlier discussion of the reasons adopters give for choosing a particular cat, we noted that many liked to believe that the cat had chosen them. This helped the volunteers as well to accept the departure of community members. If the cats indicated in some way that they were choosing their adopters, volunteers saw this as a sign that they were going home with the adopters voluntarily. It also increased the confidence of the volunteers that this would be a good adoption. Consider the following two descriptions of adoptions from our notes:

An older woman and her adult daughter have come to the shelter to adopt. They bring in a large cat carrier. Betty brings out Francis and holds him while they pet him. They are clearly looking for a big male. Betty puts him down next to the carrier and someone opens the door to the carrier. Francis walks into the

> carrier and everyone accepts this as a choice made by Francis. "Who's picking who?" says the older woman. It's settled; she will take Francis.
>
> When we come into the shelter an adoption is taking place. The story we got was that the adopter arrived early and by the time Megan arrived Murphy had hopped into the woman's lap. As Megan put it, Murphy adopted her. When Murphy was ready to leave, everyone kissed him goodbye.

Whiskers volunteers also helped cope with the disruption of adoptions by adopting the cats themselves. Adoptions by volunteers were considered the best, not only because these were dedicated cat people, but also because the shelter could thereby maintain contact with the adopted cat. Shelter volunteers were also often more willing to adopt "special needs" cats. Occasionally the shelter officers posted notices encouraging volunteers to consider adopting such cats. Glorianna was a good example of this type of volunteer adoption:

> Harriet asks me to put up a sign about Glorianna's adoption as she has many fans who will wonder where she is. Glorianna was adopted by Nelly from the Thursday morning shift. She will be an only cat and will have the run of the house, but she will spend a lot of time in Nelly's bedroom looking out the window there. This is perfect for Glorianna, who is about 11 years old. She is a pretty dilute calico who has lived in the front room cage that is for one cat ever since I can remember. This was her choice because she did not like to be out with the other cats. She has many health problems including thyroid disease. Recently, they thought she would lose an eye because

> of repeated infections, but a veterinary ophthalmologist created a little pocket to enable the eye to drain and the procedure was successful. I made the sign announcing her adoption and hung it on one of the cages where it wouldn't be missed.

Volunteers also influenced their friends and relatives to adopt Whiskers cats. These were also considered to be relatively "safe" adoptions because the shelter could maintain extended contact with the cat through the volunteer. In the following excerpt, Janet expresses her relief over the adoption of one of her favorites, Rafaella:

> The shelter is emptied out from so many adoptions. I am relieved to find that my beautiful Rafaella was adopted by the mother of one of the volunteers–Doug Andrews. He was here tonight and said that Rafaella was doing fine. She and the other cats in the house liked one another and she was already making one special friend. Her health was much improved too.

The shelter policy on returns also helped to maintain the shelter community in the face of losses through adoptions. This blanket policy was incorporated into all adoption contracts. It informed adopters that if for any reason, and at any time, they found themselves unable to care for the cat, she or he must be returned to the shelter. It also assured the adopter that the cat would be taken back, no matter what. This return policy most clearly embodied the idea that "once a Whiskers cat, always a Whiskers cat." The hope here was that the cat would be returned to Whiskers rather than being euthanized, taken to another shelter, or abandoned. One can only wonder what would have happened to Wendell if this policy had not been in effect:

> We witness a return by a mother and her adult daughter who bring the cat in. They say that the daughter's son is allergic to the cat. The mother is also allergic, but I don't have the impression she'd have returned him for herself. He was adopted as a kitten and has all sorts of digestive problems. He can't eat wet food at all and can only switch from one dry food to another very slowly. His name is Wendell and I wonder how he will survive at the shelter. He will have to be caged to maintain his diet. He is terrified, and I can't get him to look at me even when I call him by name. Then Allison who is on top on Wendell's cage sticks her head down and he turns around and sits up. They sniff each other peacefully. After the mother and daughter leave, I ask Betty what they will do with Wendell. She replies, "We had to take him back. He'll have to stay caged." She just doesn't know.

Some returns were seen as legitimate, as when the owner became ill or had to move to a nursing home. Others were considered "illegitimate," such as those that resulted from what was seen as the poor judgment of the adopters. These were usually cases where the adopter had not used common sense or had ignored the recommendations of the shelter officers:

> We learn that Minx is coming back to the shelter. Apparently the people who adopted him set him loose in a big house and Minx got frightened. Then the husband cornered him in one of the rooms and tried to pick him up and Minx bit him. Too stupid! They had been warned to confine him until he was comfortable

in his new surroundings. They were also told that he was shy and would require careful handling.

Other returns were seen as illegitimate because the excuses offered by the adopters were flimsy, self-serving, or made no sense. Amberlee, for example, was returned after three months because the family was "moving to Florida." (Are there no cats in Florida?) Returns based on feeble excuses like this, or returns that resulted from a lack of concern by the adopters, helped to unite the shelter volunteers as a special community against the often-uncaring world outside. Consider the cases of Zoe and Indigo:

> A woman comes in bringing two returns, one of whom is Zoe! She looks very unhappy. She tells Megan she can keep the carrier "as a donation."

Zoe, adopted and returned, has to readjust to shelter life.

Megan says she must sign papers giving up rights to the cat. The woman says she wants to be notified when they get adopted. She is sniffling–cold or sadness? Megan sees the woman out and when she returns we all pounce on her for the real story. The woman has allergies, Megan says, but won't take medication. She was okay with one cat but began to have trouble when she took the second one. Why, then, doesn't she keep the first one? Megan says she has been crying her eyes out for two weeks. "Who can figure?" Megan says. I say, "Well I can. These are crocodile tears designed to convince herself that she is a feeling and concerned person, which she is not."

An older woman and a younger one come in with a cat they are returning after six years! The older woman is crying. It turns out that Indigo–an all gray, good-sized male–has bladder stones and she is too cheap to pay for the surgery. So, what is she crying about? This is her choice. To boot, she says he hates other cats. So, she brings him back here–great! He is placed in the infirmary and appears very upset and bewildered. Poor guy. She is still crying when she leaves. We all tear into her as soon as she's gone.

It is ironic that one of the primary goals of the shelter, finding homes for the resident cats, also represented a threat to the shelter community. It is not surprising then that several coping mechanisms had developed to minimize disruption to the shelter community from the loss of members through cat adoptions. These ways of dealing with loss either allayed anxieties about the future well-being of the adopted cats, or they operated to maintain contact with the adopted cat.

Adoptions and the Cats Who Remained Behind

We know that adoptions affected the human–cat community, but it is more difficult to assess whether they had an impact on the cat community itself. One factor tended to minimize the kind of separations that disrupted the community. As we described in Chapter 5, the feral cats constituted the core of the cat community and they were, for the most part, unadoptable. The result was that the solid center of the cat community remained essentially intact. Although the cat community as a whole might have been unaffected by cat adoptions, the question remains, did individual cats miss those who had left the shelter? This type of loss would most likely have affected friendship pairs, but, of course, they were almost always adopted out together. We witnessed, however, the separation of one closely bonded pair of cats. The case of Sonnet and Virgo was a little unusual. First, theirs was a cross-sex friendship, whereas most of the friendship pairs we observed at the shelter were same-sex. In addition, the relationship may have been more like an adoption than a friendship as Janet noted in this excerpt:

> Sonnet has "adopted" Virgo. She is in a basket with him, wrestling with him and snuggling with him. It's pretty funny. He just loves it. Poor thing has bum tear ducts on both sides. His eyes drip and the drip stains his face. Later on the shelter is quiet. Sonnet comes over and licks Virgo, then holds him down and licks him vigorously, then wrestles with him. He seems to love it.

While Sonnet was attractive, friendly, and outgoing, Virgo was rather timid, shy, and unattractive due to his stained face. The shelter officers tried to keep them together, but the several people willing to adopt Sonnet were unwilling to take Virgo. Finally, they decided that it was unfair to deny Sonnet a home because of her relationship to Virgo, and they let her go. Virgo was caged for illness shortly after Sonnet's departure, but whether this resulted from the loss of his companion was difficult to say. He had been prone to illness before Sonnet's adoption, and there were so many factors operating at the shelter it would have been impossible to attribute it to this one event. Nevertheless, when Virgo recovered sufficiently to rejoin the community, he did become friends with the elderly Chelsea–described earlier as one of the "comforter" cats. They soon became inseparable and, once again, the shelter was considering adopting them out only as a pair. Virgo may have missed Sonnet and sought to replace that relationship, but we cannot be sure. The effects on a cat from the loss of a companion are probably better observed in a household setting. One of the shelter officers, for example, described a male cat in her home as "grieving" over the death of his father. For several days, the cat wandered around the house crying and appeared to be searching for his missing relative.

Death and the Shelter Community

Because Whiskers was a "no-kill" shelter, deaths were fairly infrequent and usually resulted from illness or old age. In 1999, the last year of our observations, we recorded seven deaths. Some of these were cats at the shelter, while others were cats who had been adopted. Usually volunteers or those close to them who adopted kept the shelter posted on the fate of their cats, such as Hobbes, of the pair Calvin and Hobbes:

> We learned today that Hobbes, who was adopted by the friend of a volunteer, has died of feline infectious peritonitis (FIP). He was so beautiful and full of life when he was here. What a shame. He was so young. His brother Calvin must miss him.

In addition, sometimes adopters who were not connected to Whiskers notified the shelter when their cat passed away as in the case of Bibbi:

> The big news today is that Bibbianne is dead. She became ill and her new owner took her to her veterinarian. After examining her, he said there were so many things wrong with her, including a thyroid condition, that he could not make her well. They decided to put her down. Betty was crying in the telling of the story that she got second hand, I think from Megan. I wonder if the woman sought Megan's approval since she was a new owner. Now Lisa [Bibbi's friend who was adopted with her] is alone. The woman said she would choose a companion when she was less upset.

Given the shelter's belief in the uniqueness of each cat, and that all Whiskers cats remained in the community, all deaths were taken seriously and news of deaths spread quickly. In the excerpt from our notes below, we learned of the death of Philip, one of the long-term residents of the shelter. Philip, remember, was a feral cat who bared his teeth and hissed if someone got too close, but several volunteers had set their sights on bringing him around and were occasionally successful in petting him from behind:

> Kate tells me that they had to put Philip down today. He's been sick for such a long time with multiple

> problems from mouth ulcers to liver disease. She said he was too ill to go on and, of course, being feral, very difficult to treat. He had been at the vet for a while. He was 10 to 12 years old.

The death of cats who were particular favorites among the volunteers or who had been long-time residents at the shelter might result in the posting of a special "death notice." One such notice was for Geraldine, the small black cat who had a disorder that made her unable to control her movements. She frequently fell over and was incontinent. All of the volunteers loved her, and the cats were generally tolerant of her efforts to snuggle with them. The notice read:

> Our precious Geraldine lost her valiant struggle Saturday. Megan, Kate and Betty spent her last hours with her at ___________ Veterinary Hospital. Unfortunately, we could not help her and could not stop the disease process. Rest well, little one. We'll all miss you.

Wesley also received a posted death notice. Wesley was a very old cat who had been found unconscious in the middle of a country road. He became a favorite of the shelter officers and volunteers, and one volunteer eventually adopted him. His death notice read as follows:

> Our Wesley passed on–he left us on January 29th. Our sincere thanks to Diane Morse for giving him a loving home.

Although the shelter made extraordinary efforts to preserve life, it did, as in the case of Philip, turn to euthanasia when cats were suffering and nothing more could be done for them medically. It was never easy for shelter staff to decide when to end the life of a sick cat, as seen in the case of Dot, a white semiferal cat who had been at the shelter for several years:

> Dot was caged as she has been for some time. She began sort of gagging and pulling at her mouth with her paw. Betty and I went over to the cage. Kate, who had been observing Dot, called Harriet asking her to have Dotty euthanized because this had been going on for months. The doctors could not find out what was wrong and Dotty was in terrible pain. Meanwhile, Dotty began eating and seemed better. I felt quite uncomfortable with the decision to euthanize since she felt better.
>
> Dot was euthanized last Monday. Betty told me when she called to tell me that Rosie's adoption was going well. Apparently, when Harriet came to get Dot in the morning she was screaming in pain and tearing at her mouth again. I guess nothing more could be done.

From time to time there were "natural" deaths at the shelter. These were most likely to occur among the more reclusive feral cats whose illnesses might go undetected by the volunteers. One of the more distressing examples of this was the case of Misti, mentioned earlier. She was an extremely shy gray tiger. We first noticed her in the kitchen at feeding time. She was always behind the refrigerator and would cautiously creep over to the nearest food plate to grab a few bites when no one was there. When Laura fed the cats, she always put a small plate next to the refrigerator so Misti would be sure to eat. It was Laura who gave us the news of Misti's death:

> Laura reveals that she was very upset last week when Misti was found dead under the refrigerator. Laura said that she was very decayed, indicating that she must have been dead about two weeks. That was the last time anyone remembers seeing her. Laura thought it meant the volunteers were not moving the refrigerator, which is on rollers, to clean as they should be doing. Betty said she was devastated when she heard about Misti and that it would be necessary to devise a system for checking on ferals. She said that Harriet was working on it. Laura said she went home and gave her own kitties a special hug. I talked with Betty about the possibility of everyone taking responsibility for observing one or two ferals every time they come to the shelter. She thinks it's a good idea. Steve and I pick ours and put our names next to their photos on the feral board.

Overall, the loss of cats through death did not cause a major disruption to the human–cat community. To begin, the death rate of Whiskers cats was relatively low. Add to that the fact that some portion of these deaths came at the end of a long and happy life–either at the shelter or in the home of an adopter. Although there was sadness, there was also the satisfaction of having rescued and, in many cases, found homes for the cats. These were the success stories.

The cat deaths that most disrupted the community were the untimely deaths–those that resulted from incurable diseases such as feline leukemia and feline AIDS or other diseases that could not be identified and treated. The shelter made every effort to find suitable homes for those cats who tested positive for these diseases but who were not yet showing symptoms. Misti's death, described above, was probably the most traumatic for the shelter because, to some degree, it represented a failure of the community to look after its own. That the shelter immediately took steps to implement procedures to prevent this happening in the future was evidence of the volunteers' distress.

The impact of cat deaths on the cat community was probably minimal because of the relatively low number of deaths and because the cats rarely confronted the death of their coresidents directly. We did wonder whether the death of a cat like Philip, who was a core member of the feral community, disturbed the cat community in some way, but we have no direct evidence of this. Having come full circle now, from entering to leaving the shelter community, in the next chapter we reflect on what we have found.

7

Culture and Self in the Domestic Cat

MARIAN STAMP DAWKINS,[1] and most other biologically trained scientists, strongly believe in Occam's Razor, also referred to as the principle of parsimony. This principle basically states, "We should always start with the simplest explanation and only when this has been shown to be quite inadequate should we move on to a more complex one."[2] The problem is that many scientists do not move on, no matter what the evidence, leading to a strong reductionist preference in science. This has been especially true in the study of animals, where many scientists still cling to behaviorism in spite of almost 20 years of research undermining their position.

Sociologists, from the beginning, have been no friends of parsimony. Instead they have argued in favor of the principle of emergence. No one articulated this more cogently than Emile Durkheim, who saw social and cultural forces as creating a level of analysis and explanation distinct from biological and psychological forces:

> To be sure, it is . . . true that society has no other active forces than individuals; but individuals by

> combining form a psychical existence of a new species, which consequently has its own manner of thinking and feeling. Of course the elementary qualities of which the social fact consists are present in germ in individual minds. But the social fact emerges from them only when they have been transformed by association since it is only then that it appears. Association itself is also an active factor productive of special effects. In itself it is therefore something new. When the consciousness of individuals, instead of remaining isolated, becomes grouped and combined, something in the world has been altered.[3]

Thus, social phenomena are no mere metaphors reducible to biological or psychological entities. Rather, they have explanatory powers in their own right, as Durkheim demonstrates for the case of suicide in his famous work of that name.

Durkheim's view remains dominant among sociologists even though sociobiology has been mainstreamed to some extent judging by its inclusion in theory textbooks and even in some introductory sociology texts. Many sociologists, nevertheless, see the ideas of sociobiologists as supporting ultraconservative stances on major social issues:

> This approach met with enormous hostility from some sociologists and, indeed, continues to do so, largely because it appears to legitimate aspects of human societies that people wish to reform and to set limits on how far you can change . . . people. . . . They argue that Wilson, for example, is a product of an alienated culture and his own class prejudices, and he "joins a long parade of biological determinists whose work has served to buttress the institutions of

> their society by exonerating them from responsibility for social problems." The question of what is biologically . . . based becomes especially controversial in the context of discussions of gender roles and the origin of differences between men and women; homosexuality . . . and the extent to which intelligence is inherited or the product of upbringing.[4]

Sociologists, then, generally continue to see humans as largely shaped by social and cultural factors, with biology playing a limited role in providing a context within which society and culture vary. Even those social and cultural factors are not determinative for symbolic interactionists who argue that humans are guided not only by norms and rules from their culture but also by their own goals and the immediate situation in which action takes place. This means that the courses of action chosen by social actors are negotiated and variable. Small groups may differ from the whole on this basis, and individuals may differ from the group as well.[5]

Thus, sociologists have historically been hostile to reductionism, arguing that social phenomena can only be explained in terms of other social phenomena and not in terms of biological or psychological variables. Further, they have rejected efforts to find any single variable or limited set of variables even on the social and cultural levels that can explain social facts that are instead seen as complex and multicausal. Hence, in our approach to the study of the social relationships of animals we are likely to be more skeptical of the value of parsimony than our colleagues in biology. We believe our skepticism has been borne out by the research on animals accomplished in the last 15 to 20 years. These works include all those cited in Chapter 4 that suggest the need for more complex theories of animal behavior than have been presented in the past. Further, there are few better examples of nature's lack of parsimony than the case of domestic cats. Domestic cats have a social capacity that far exceeds what they have been able to display in most places and at most times.

The Cat Community at the Shelter

Based on our own study of the Whiskers shelter, we strongly support the need for more complex theories of cat behavior. Looking first at the social structure of the cat community, we found an almost complete absence of either a hierarchical order or dominance behavior. It is possible to create hierarchy in cat groupings in the laboratory through extreme overcrowding and limited access to food. This results in the emergence of both dominant and pariah cats.[6] One also finds hierarchy in natural groupings of feral cats. Among the breeding males there is usually one who monopolizes sexual access to a group of females by fighting or intimidating other males.[7] Among the females in feral colonies, kinship plays a role in hierarchy. Tabor, who observed feral cat colonies, noted that other cats granted the matriarch or senior female in the group special deference when the cats were being fed by a kindly lady. They would give the matriarch first choice of the food plates insofar as "she was aunt, mother or grandmother to most of those present."[8]

With the exception of overcrowding, none of the conditions described above were present at the Whiskers shelter. All of the cats were spayed or neutered, which eliminated sexual competition among them. Food was plentiful, which eliminated competition over eating, and kinship among the resident cats was relatively rare. These conditions, then, gave the cats the freedom to shape their community as they desired. As a result, we found an egalitarian community in which the absence of competitive pressures, in combination with the very crowded quarters at the shelter, brought out the social needs and capacities of the cats. In other words, the shelter conditions both freed the cats, and provided the opportunity, to pursue, in Maslow's[9] terms, their higher need for sociality.

The cats' efforts to satisfy their social needs were reflected in the social structure of the cat community. To begin, when we observed the patterns of association among the cats, we noticed an elaborate network of overlapping friendships centered around the old-timers at the shelter. While a few of the cats maintained an exclusive relationship with another cat, the more usual pattern was to have multiple friends with whom one spent time, played, and slept. Thus, close, nonkin relationships, including some opposite-sex friendships, were not only possible, but also entirely unproblematic for the cats. These friendship networks held the cat community together and formed the basis for its social cohesion.

The satisfaction of social needs among the cats was also supported by a set of norms and sanctions that were part of the social order and culture of the cat community. The norms encouraged friendship, physical closeness, and the sharing of space, and they discouraged aggression, territoriality, and dominance behavior. Both the cats and the volunteers sanctioned conformity to these norms. Cat sanctions tended to be positive in that those cats who did not show aggression or dominance behavior were rewarded with affection and friendship; they had partners with whom to play and buddies with whom to snuggle and sleep. We did occasionally observe negative sanctioning on the part of Marquis, the shelter guardian, who would often move in to break up altercations between other cats.

For their part, the volunteers offered both positive and negative sanctions. On the positive side, they allowed the healthy cats to freely associate with one another, rather than caging them. They also provided several soft places where the cats could snuggle, groom each other, and sleep together. These were the blankets on the cage tops, cat beds, and baskets. And, on occasion, volunteers would maintain special places for friends, such as the litter pan for Bibbi and Lisa described earlier. On the negative side, although this happened rarely, the volunteers temporarily removed aggressive or disruptive cats from the free population.

As described earlier, the cat culture we found at the Whiskers shelter emphasized friendship, affection, and social cohesion. By culture, we mean the particular adaptations the cats made to their living conditions at the shelter, which were transmitted to the new cats entering the shelter. One of the striking features of this culture was the ready sharing of space and the high tolerance for physical closeness. Territoriality, or the defending of space, was essentially absent in the shelter, suggesting that domestic cats are not territorial by nature, although they will form territories when it helps them survive.

The feral cats played a central role in the creation and maintenance of the culture we are describing. They were the old-timers at the shelter, and many had come from feral colonies and were used to forming relationships with other cats. The feral cats transmitted the norms of this culture to the new residents as they were first exposed to, and then released into, the free population of cats. Initially, many of the new cats were hostile, hissing at and swatting anyone who came near them. Then, with few exceptions, they calmed down and peacefully assimilated into the cat community.

The gradual manner in which the volunteers introduced the cats into the free population greatly facilitated their acculturation. However, we believe that the primary mechanism of cultural transmission was the positive sanctions that supported the norms of sociability at the shelter. Once a cat accepted these norms and put aside his or her hostility, all of the benefits of community became available: friendships, grooming and cuddling partners, and the comfort and warmth of the shared soft places in the shelter. Thus, although the cat culture we found at the shelter was supported, and indeed made possible, by the policies and practices of the volunteers, this culture ultimately emerged out of, and was sustained by, the interactions among the cats themselves.

We add one final note to this discussion of the shelter cat culture: the importance and symbolic meaning of place. We have described how, with our own cats, certain places in our house came to have a shared meaning beyond any immediate function these places served. When we or Jenny went to a certain throw rug upstairs, it meant it was time to play with a string or a ball, and when we or Annie went to a certain radiator cover, it meant it was okay to snuggle and smooch. We believe this same phenomenon occurred among the cats themselves at the shelter with regard to the cage tops, beds, and baskets. Certainly for the individual cats, these were comfortable places to relax and sleep, but it was the ready and peaceful sharing of these places that gave them a higher meaning and made them symbolic of the social integration of the cat community.

The Social Self in the Domestic Cat

In Chapter 3, we began with a discussion of the nature of the self in humans and pointed out that the existing data supported Charles H. Cooley's theory of the self for nonhuman animals as well. The most social apes had the strongest sense of self. This was true whether they were social with one another or with humans. Conversely, those who were isolated from social contacts had the weakest sense of self. We argued, thus, that it was as fruitless to look for the self in the biology of cats as it was to look for it in the biology of humans. Rather, our efforts should be focused on the social conditions needed for self-enhancement and development. We found a number of these conditions at work in the shelter in the relationship between the human volunteers and the cats. The volunteers gave every cat who entered the shelter a name and constructed life histories for every cat. They noted and passed on to other volunteers the observed likes and dislikes of the cats. Then they fostered the cats' preferences for food or friendship choices, for example. They tried to discern the feelings and motives of the shelter cats and took those feelings seriously, particularly in the case of depression. They interacted with the cats based on how they perceived the cats were feeling, just as one would with any friend.

The overwhelming majority of the cats were highly interactive either with the volunteers, one another, or both. That is, they frequently initiated social encounters with others in the course of a day. These encounters strongly suggested a well-developed sense of self. We particularly point out the intimacy of the cage tops and sleeping spots, the many instances of empathy demonstrated by the free cats for the caged cats, and, of course, the friendship patterns of the shelter. Since few of the shelter cats were related to one another biologically, it is difficult to escape the conclusion that their completely voluntary friendship choices were expressions of preferences for others of their kind with whom they had a special rapport or affinity. This could only be acquired through taking the role of the other, which is the central ability required in the emergence of a sense of self.

Another characteristic of the shelter cat and the cats in the multi-cat households we studied is their highly developed individuality or distinct personalities. Obviously, there are similarities in cat self-expression as there are in human self-expression because each species shares the same anatomical, facial, and other characteristics. But the way in which each cat combines them to communicate with others is highly unique. Marquis was a wonderful example of the uniqueness of the shelter cat, with his self-initiated security guard activity and rejections of the homes that were found for him. But the shelter was full of unique cats, as were the multi-cat households. No two of the fifteen cats who have lived with us over the years were alike. Each has or had special ways of expressing himself or herself. For instance, when Calvin wants to get our attention, he collects all of his little wool-like balls and brings them to us. Often he vocalizes as he looks for them. When he first started to do this, he wanted us to play ball with him. Over time, however, the balls came to indicate other desires. Most often he just wants attention–to be petted and scratched. Sometimes he wants food. Sometimes he wants to play with something other than the balls, such as a mouse. Thus the balls have become a symbol for a state of mind. None of our other cats seeks our attention in this way.

The evidence that cats have a sense of self seems to us compelling. We believe that the strongest sense of self is found in cats who receive a lot of attention from humans, other cats, or both. Like the human self, the cat self is a product of intimate, positive social interaction. That attention is most likely to be given by "cat people," who are devoted to their cats and cat colonies, such as that formed in the shelter.

The shelter cats, in fact, meet all of Dawkins's criteria of self-awareness. One final point worth making is that like humans, cats develop, learn, and grow emotionally over time. They are not static beings; the events of their lives change them. To know this one must observe and work with them over a significant period of time. Kate's success with feral cats, for instance, came after a lengthy investment in time. Such change and growth cannot be discerned by the experimental method in which an animal's behavior may be observed for 15-minute intervals.

Understanding Other Species

We agree with Dawkins that choice is an important tool in assessing the needs of other species.[10] The choices the Whiskers cats made enabled us to learn many things about them. Researchers who seek to understand other species must incorporate choices into their studies. We must also go further. It is time we weighed in with the evidence and stopped asking the same questions repeatedly: "Are animals conscious?" "Do animals feel?" or "Do animals have cultures?" Instead, at least as regards mammals and birds, we must move forward and begin to ask how consciousness, feelings, and culture affect animal lives. Our own study generates a number of new questions for future research: What are the social conditions that foster self-awareness in animals? What are the social conditions that nurture affection and social cohesion, as opposed to dominance and aggression, in animal groups? How does territoriality vary with different social and physical conditions? How do animal cultures develop and affect individual behavior? How do the feelings of nonhuman animals motivate them, and how can we best understand those feelings? We must put aside the myths about felines and engage in real observations of real animals.

We further argue that the ethnographic method as used by sociologists is an essential tool in understanding other species. Like traditional ethology developed by biologists, using this method allows the researcher to observe animals in a natural setting in which one is more likely to find typical behavior. It allows for observation of the same animals over a long period of time and it makes possible the study of the growth and development of individual animals over time. Because of the leisurely nature of such studies, a researcher can observe the same phenomenon repeatedly and refine and adjust his or her interpretation and perspective. Ethnography, however, goes beyond traditional ethology in two important respects. First, ethnographers attempt to take the viewpoint of the animals under study and see things from their perspective. Bekoff has recently urged biologists to take this approach, which he calls "deep ethology,"[11] so it may be that some convergence of the two methods is possible. Second, ethnographers seek to interact with, and participate in the life of, the animals they are studying. This is the extension of participant observation in sociology to include human–animal interaction. By using the ethnographic approach, a researcher can know his or her subjects far more intimately and extensively than with other methods. One final note: Although the ethnographic method is qualitative, it does not preclude quantifying when appropriate.

For instance, it was valuable to us to count the number of times particular cats slept with each other, although the figures could not be considered final given that we did not live at the shelter.

We cannot continue to rely exclusively on the experimental method in which a small number of animals are observed for short periods of time to understand other species. We have been terribly misguided by the general failure of experimentalists to study key subjective phenomena such as self-awareness. And when they do attempt such studies, the animals are sometimes manipulated in highly contrived and convoluted situations. Often, as in the Gallup studies cited in Chapter 3, generalizations are made to entire species based on very small numbers, and variations within species are underexamined or even ignored.

Implications of Our Findings for Social Policy

The domestic cat is not limited to one nature. Rather, cats carry within themselves many possibilities, which may or may not be realized in a particular setting. They can survive alone if they have to, but all of the evidence from the shelter indicates that a solitary life is often hazardous. When such cats arrive at the shelter their health usually has been badly compromised. They have a great social capacity that emerges under the right conditions. Under those conditions they demonstrate a strong need for affection and friendship and will form socially cohesive groups. These findings, we believe, have important implications for social policy.

As companion animals, cats have certain legal protections in many states. For example, companion animals may not be eaten. In New York state since 1998, drivers who injure a cat must try to locate the animal's owner or notify police to avoid a fine.[12] That same state passed a bill in June 1999 making acts of aggravated animal cruelty felonies, which carry jail sentences, rather than misdemeanors, which just impose fines. The law was restricted to companion animals defined as dogs, cats, and "any other domesticated animal normally maintained in or near the household of the owner or person who cares for such dog, cat or other domesticated animal."[13] As of 1996, four states had laws restricting experiments on living animals in public schools grades K–12. These restrictions included mammals and birds in New York and Florida, vertebrate and invertebrate animals in California, and vertebrate animals in Pennsylvania. Although the law was not framed only in terms of companion animals, they benefited from it.[14]

In spite of these protections for companion animals, nearly 5 million dogs and cats are euthanized each year in shelters that have exceeded their housing capacity.[15] In 1998, 24,712 cats and 76,071 dogs were used in labs in the United States,[16] virtually all of whom are killed after the experiments are completed. An additional unknown number die each year from undetected abuse and neglect. They are discarded like old clothes when their owners become tired of them or find them inconvenient. Thus, the legal protections they currently have do not spare them many types of suffering. As aware creatures, these various forms of mistreatment matter to them.[17] They become depressed, fearful, listless, and the light goes out of their eyes, as it did for Nathan and Amberlee at the shelter when their owners returned them.

The mistreatment of animals matters to us as well. Because we are animal activists as well as social scientists, the findings of our study have made us even more committed to the animal rights point of view. Every level of government needs to recognize its responsibility to the animals in its jurisdiction. The federal government should create more uniform, national legislation to protect companion animals, because the states vary enormously in their legal protections.[18] Federal legislation should also severely restrict companion animal use in scientific experiments and education. In addition, federal laws could assist states in restricting companion animal breeding to control population size, as many breeders ship animals across state lines.

States must also pass legislation on these issues. For instance, legislators could mandate educational programs in the schools on responsible pet ownership, which some states are already doing.[19] They could restrict breeding of companion animals. They could also strengthen any existing legislation by funding strong enforcement measures and providing grants to no-kill shelters to spay or neuter before adoption and expand and improve their facilities, which would then be inspected as well. Municipal officials also must take responsibility for the companion animals in their jurisdiction. They need to subsidize spay/neuter programs for pets as well as spay/neuter/vaccinate and release programs for ferals. They could then work with shelter staff and community members to provide adequate nourishment for ferals, who perform a public service by keeping rodent populations under control. They also should subsidize enlarging and improving shelter facilities.

Such measures would greatly relieve the suffering of companion animals. Also, the definition of companion animals should be expanded as more is understood about the attributes of nonhumans and their value in human–animal relationships. California residents did so in 1998 when they passed the criminal law prohibiting the slaughter of horses and the sale of horsemeat for human consumption. The advocates of this legislation successfully argued that horses were recreational animals, a part of the heritage of the state, and they should be legally protected from slaughter, just as were dogs and cats.

Unfortunately, companion animal suffering is not only increased by inadequate legislation but also by existing conventions, administrative rulings, and laws that make it difficult to care for and about one's animals. For example, local officials prevent owners from rescuing their animals in natural and human-made disasters that require evacuation. Even owners willing to sign waivers releasing officials from liability if owners are injured are not allowed to return to their homes to evacuate their animals when there are risks associated with the disaster at hand. If owners are with their animals and must seek public shelter in such a situation, the shelters do not have facilities for animals and will deny them entrance.

Overall, we argue that the issue is one of inclusion in the social bond. We have not yet reached the stage in which we consider our companion animals part of our community with a right to moral consideration. But as creatures who enter into complex social relationships, are able to see things from another's point of view, learn from one another, adapt to novel circumstances, and indicate self-awareness and individuality, we believe that such social inclusion is their due.[20]

8

Animals in the Future of Sociology

WE LOCATE our ethnographic study of the Whiskers Cat Shelter in the rapidly growing subfield of sociology, society and animals. Sociologists in this subfield recognize the many roles that animals play in all societies–as companions, sources of food and clothing, subjects of medical and behavioral research, participants in sports, entertainers, and wildlife. Animals influence virtually every aspect of our lives, which is reflected in the growing body of research in this area.

We found a broad range of topics and approaches in the journal, *Society and Animals,* over the past several years. Researchers in several studies looked at factors that shape our social constructions of animals and how these constructions can affect our treatment of, and our relationship to, these animals. Scarce,[1] for example, examined how economic considerations and public policy shape scientists' constructions of salmon as more than just an element of nature. Wolch, Gullo, and Lassiter[2] analyzed the print media to document changing conceptions of cougars over time in California and how this has affected their legal status (i.e., whether they can be hunted). And Quinn[3] looked at the way in which paintings and drawings of idealized animals shape standards for particular breeds of domestic cattle. Researchers have also applied the perspective of social constructionism to studies of social problems relating to animals. Mauro,[4] for example, studied the efforts of an animal rights group in Australia to redefine duck shooting as animal cruelty and, therefore, as a social problem to be addressed. Likewise, Carbone[5] examined the history of the debate over whether the employment of rodent guillotines to decapitate research mice is sufficiently painful to define this practice as a social problem in need of remedy.

Researchers who study social movements have found campaigns on behalf of animal rights to be a fruitful area in which to apply and expand their theories and perspectives. Galvin and Herzog,[6] for example, surveyed participants in the 1996 March for the Animals in Washington, DC, to ascertain their views on various aspects of the struggle for animal rights. Taking a different approach, Munro[7] focused on the animal rights movement to illustrate the importance of studying countermovements or movements that defended such targets of animal activists as factory farming, vivisection, and recreational hunting. Other researchers represented in this journal explored the links between animal abuse and domestic violence. Ascione, Weber, and Wood[8] looked for this link in their survey of shelters for battered women, and Raupp, Barlow, and Oliver[9] looked for the link in the childhood recollections of the college students they surveyed. Taking a somewhat broader approach, Nibert[10] surveyed adults in Ohio and found that support for animal rights was linked to support for several human rights issues. In other words, animal people were people people too.

Finally, researchers in two studies[11] addressed the issue of species boundaries. They looked at the phenomenon of transplanting human genes into animal bodies and animal organs into human bodies, and suggest that these medical procedures have a variety of social, cultural, and psychological ramifications. They erode the traditional boundaries between humans and animals and challenge our definitions of human and animal. Based on this brief sampling of articles from *Society and Animals*, then, we can see that human–animal relationships offer a wide range of opportunities to apply and refine sociological theories and perspectives.

The Shelter Study's Contribution to the Study of Human–Animal Relations

Although our study has theoretical links to other work on human–animal interaction, we make a different contribution than most of the studies done to date. For one, we study human–animal interaction in which the animals are not subjects but partners in the interaction. That is, at Whiskers, the cats help to create the structures and culture of the shelter. Although they are *our* subjects, along with the humans, we try to understand and present their perspective on their relationships with people. In this sense, our study is linked to the other work in this field because this aspect of it is squarely in the social constructionist framework. The human volunteers and the resident cats at Whiskers construct their world through an interaction process. What is new is taking the animal's perspective into account.

A second focus that is unique is our attempt to show that, just as in the case of humans, the self in animals is social in its nature and development. That is, although there is surely a neurological basis for the self, it only emerges through intimate interaction with others. This argument flows from the work of Cooley[12] and Bogdan and Taylor[13] in sociology as well as from that of animal scientists such as Suarez and Gallup.[14] Using this conceptualization of the self, we examined the role of the volunteers in bringing out and enhancing the social self of domestic cats through their intimate interactions at the shelter.

A third contribution of our study is that we examined the role of emotions in the human–cat community of the shelter. First, the volunteers provided a safe environment in which the cats could feel secure and comfortable. Within that context, they discouraged negative emotions that might lead to violence and they encouraged and provided opportunities for the expression of positive emotions. The resulting emotional climate both enhanced human–cat relationships and also nurtured the cats' individual self-development. The cats felt free to explore social goals and did so with both humans and other cats. There is a biological basis for these findings. According to the polyvagal theory, the mammalian nervous system has evolved in such a way that it can react to challenges to survival through communication via facial expressions and vocalizations. At the same time mammals retain earlier means of coping such as "fight or flight" on the one hand, or "play dead" on the other. Whether mammals use the more sophisticated system of "talking one's way out of trouble" or the earlier systems depends upon the safety of the environment. The use of communication to solve problems is more sophisticated because it allows for a great variety of positive social behaviors:

> To survive, mammals must determine friend from foe, evaluate whether the environment is safe, and communicate with their social unit. These survival-

> related behaviors are associated with specific neurobehavioral states that limit the extent to which a mammal can be physically approached and whether the mammal can communicate or establish new coalitions. Thus, environmental context can influence neurobehavioral state, and neurobehavioral state can limit a mammal's ability to deal with the environmental challenge. This knowledge of how the mammalian nervous system changes neurobehavioral states to adapt to challenge provides us with an opportunity to design living environments, which may foster both the expression of positive social behavior and physical health.[15]

For the resident cats at Whiskers, the shelter was such a safe environment, allowing for the pursuit of social goals in relation to humans who, as a different species, constitute a significant challenge. It also allowed for the pursuit of social goals with one another.

Thus, the most unusual aspect of our study is that we used the methods and theoretical perspectives of sociology to examine the social and cultural lives of animals themselves. This has convinced us that sociology can make a distinct contribution to the study of animal life, both theoretically and methodologically.

Sociological Theory and Method in the Study of Animal Life

In the course of our research, we have borrowed heavily from the symbolic interactionist perspective, using the ideas of Cooley,[16] Mead,[17] and Collins.[18] The essence of symbolic interaction involves taking the role of the other, defining situations, selecting courses of action based on those definitions, and producing shared meanings with others, which, sometimes, transform objects. These shared meanings are the rudiments of culture. They are also the essence of the self that forms as one is able to see oneself as an object as seen by others. In Mead, such symbolic interaction can only occur through language. One must be able to converse with oneself covertly in words and sentences before acting, and the meanings shared with others are largely linguistic meanings. Language itself is a transformation of objects, in that the word comes to stand for the object. This is Mead's essential mistake. Language is not the basis of symbolic interaction, and sociologists must begin to update Mead's ideas.

Interestingly, Cooley knew this,[19] and the research of modern psychology and, particularly, neurobiology has borne him out. For instance, the neurobiologist Antonio Damasio[20] argued that consciousness is rooted in the representation of the body. The organism is represented in its own brain for monitoring purposes. He calls this the proto-self and it is not conscious. The body and other objects are represented in the brain in images. Action is tied to the availability of such guiding images in the brain, which allow us to choose among previously available patterns of action in order to increase the probability of the success of the action. From this perspective, consciousness is a device to allow for the effective manipulation of images in the service of the organism. Actions can now be shaped by a concern for one's own life. Thus consciousness is knowledge. It is a wordless knowledge. It is a *feeling* of knowing when one is processing an object:

> Consciousness begins when brains acquire the power . . . of telling a story without words, the story that there is life tick-

> ing away in an organism, and that the states of the living organism . . . are continually being altered by encounters with objects or events in its environment . . . or by thoughts and internal adjustments of the life process. Consciousness emerges when this primordial story–the story of an object causally changing the state of the body–can be told using the universal nonverbal vocabulary of body signals. The apparent self emerges as the feeling of a feeling. . . . I suspect consciousness prevailed in evolution because knowing the feelings caused by emotions was so indispensable for the art of life, and because the art of life has been such a success in the history of nature.[21]

Consciousness and emotions prevail together. Emotions cause feeling states; to know what one feels motivates action.

Damasio goes on to distinguish two types of consciousness: core consciousness and extended consciousness. Core consciousness involves the nonverbal awareness of one's existence and state of being. It involves only short-term memory. Mostly it exists in the here and now and is stable across the lifetime of the individual. It is widely distributed throughout the animal kingdom and is the foundation upon which extended consciousness develops. Extended consciousness encompasses varying levels, allowing for the emergence of a longer-term memory and a sense of the future. It develops through the lifetime of the individual, permitting the emergence of an autobiographical self. It also is found among the higher organisms in the animal kingdom and reaches its full flowering in humans who add language to the repertoire of consciousness-enhancing devices. When core consciousness is damaged, extended consciousness does not survive:

> Language–that is, words and sentences–is a translation of something else, a conversion from nonlinguistic images which stand for entities, events, relationships, and inferences. If language operates for the self and for consciousness in the same way that it operates for everything else, that is, by symbolizing in words and sentences what exists first in a nonverbal form, then there must be a nonverbal self and a nonverbal knowing for which the words "I" or "me" or the phrase "I know" are the appropriate translations, in any language. I believe it is legitimate to take the phrase "I know" and deduce from it the presence of a nonverbal image of knowing centered on a self that precedes and motivates that verbal phrase. . . . The idea that self and consciousness would emerge *after* language, and would be a direct construction of language, is not likely to be correct.[22]

Thus, the sense of self is founded on images stored in the brain and is not confined to humans or, even more narrowly, to humans with language. The ability to take the role of the other, define situations, produce shared meanings, and transform objects into meaningful symbols, then, is not founded on language. As a consequence of such findings, it will be necessary for sociologists to rethink the whole connection between language and interaction in relation to humans.[23] We must also rethink the exclusive attention we have devoted to the study of humans and consider broadening that attention to human–animal interactions and to relationships among animals. In short, these findings of neurobiology have profound implications for future directions in sociology.

Returning to the implications for our own study, we note that in Damasio's terms domestic cats would certainly have core consciousness. We believe that our own evidence indicates that they also have some degree of extended consciousness. They remember things for a considerable period of time, their past experiences affect their present actions, and they are capable of fairly sophisticated relationships with their own kind and with humans. Thus, feral cats were able to use their past experiences in cat colonies in forming new relationships at the shelter, which represented a novel solution to the problems of overcrowding they faced there. The shelter cats also formed relationships with the many humans they encountered at the shelter and these relationships were products of interaction, and differed from each other. Thus, the study of domestic cats and other such animals would fall well within the range of sociological endeavor. With the appropriate revisions to symbolic interactionism, we have an important theoretical perspective within which to consider such interspecies and within-species relationships.

As part of these revisions, we must increase our emphasis on the study of nonverbal behavior. The ethnographic method provides a ready tool for such analysis in that one can as easily observe facial and body expressions as one can words. The use of video cameras would also be a big boost to the study of nonverbal behavior; we now wish we had used the camera to a greater extent in our own study.

We can also apply what we found regarding the role of emotions at the shelter, which we described earlier, to the sociology of emotions. The fact that the volunteers acted in ways designed to shape the emotions of the cats in a particular direction supported McCarthy's claim that emotions are "cultural acquisitions."[24] That is, the peaceful and cohesive cat community we found at the shelter was, at least in part, the outcome of its cultural practices. Our study can be seen, then, as extending the analysis of the role of culture in shaping emotions to animal emotions in a human–animal community. That humans and cats are able to read each other's emotions is based on the fact that they, and indeed all mammals, share the same biological basis for emotions. Emotional responses, according to Damasio,[25] have evolved over time and "are part of the bioregulatory devices with which we come equipped to survive."[26] These evolutionary similarities regarding the capacity for emotion provide the basis for interspecies communication between humans and animals and among animal themselves. Those who have studied the relationships between predators and prey, for example, have found that survival of both depends on their ability to read the emotions and, therefore, the intentions of the other.[27] The fact that humans and animals can form successfully functioning communities, such as the shelter community we observed, also testifies to the capacity for interspecies understanding. Given this, there is no reason why sociologists cannot undertake, and gain insight from, the study of animal emotions.

One of the more significant aspects of our study was that we discovered a distinctive cat culture at the shelter. Supported by the practices and policies of the volunteers, this culture emerged out of the interaction among the cats and was maintained and transmitted to new cats by the mostly positive sanctioning of the cats themselves. In seeking some perspective on animal culture we had to turn to the work of biologists, psychologists, and anthropologists since sociologists have yet to address this well-established phenomenon. Sociologists must study animal culture, not only because culture is central to sociology, but also because it would help sociologists to transcend the legacy of Mead and stop using (and misrepresenting) animals as the opposite of humans.

Finally, we stress that sociologists must extend the perspective of social stratification to the study of human–animal relations. Animal rights advocates see that animals are severely oppressed in modern, industrialized cultures. Although we did not focus on animal oppression in our study, since we studied a shelter, we did emphasize the rescue efforts that are set in motion by such oppression. Just as there are shelters for dogs and cats, there are shelters for farm animals, horses, and wild animals who suffer greatly in our society.

To address how our society oppresses animals and how sociologists should confront such issues, we must first examine some of our cultural practices. To begin, we eat a fair range of animals. According to Singer, "Over 100 million cows, pigs, and sheep are raised and slaughtered in the United States alone each year; and for poultry the figure is a staggering 5 billion."[28] Most of these animals spend their short lives in the most abysmal of conditions under the factory farm system in which they are confined to small crates or tied to a post in such a way that they may not even be able to turn around. They are forced to breed, and pregnant sows are locked in small gestation crates where they can find no comfort. When their babies are born, the sow has no access to them. They, in turn, are only allowed to reach the mother's teats and are stressfully weaned at three weeks, which is far too early. Then they are plied with hormones to make them grow faster and antibiotics to enable them to live long enough to reach market.[29]

We also experiment on them. It is very difficult to obtain total figures on use of animals in experiments. We were able to give figures for dogs and cats in Chapter 7 because they are among the privileged species that the U.S. Food and Drug Administration (USDA) counts. Rats and ferrets, who are not counted, make up the majority of the animals used. Although the numbers of dogs and cats used in research has declined, there has been an increase in the use of species the public cares less about and who are not counted.[30]

Next, we dispose of animals we do not want as if they were consumer items no different from a toaster or a pair of unfashionable shoes. The figures on euthanasia in shelters were given in Chapter 7. We were only able to conduct our study because of people's consumerist attitude toward animals, which resulted in the formation of Whiskers and the rescue of many abandoned cats. These consumerist attitudes are closely related to issues of animal cruelty, which is widespread and in need of sociological investigation.

Finally, wild animals who perform in circuses and other entertainment venues often are confined to small cages and are beaten to make them conform to the trainer's wishes. These conditions are imposed on large animals such as elephants and tigers, who would normally roam over an extensive territory. Sometimes they express their tension by attacking their trainers or other humans.

We believe it is incumbent upon sociologists to begin to explore these issues from the perspective of stratification and oppression. Sociologists have successfully used this perspective to examine the condition of minorities and women and have provided numerous studies, analyses, and statistics that have helped shape public policy in a less discriminatory and more protective direction. We can do the same thing for animals. A beginning has been made, not so much in the areas mentioned above, but in the area of animal abuse and its relationship to human abuse. Researchers have carried out several studies, a couple of them noted above, that focus on animal abuse and domestic violence. They have explored the question of the extent to which the two are linked. As a result of such studies, officials have begun taking animal abuse more seriously, and legislators have passed laws imposing stiffer penalties on animal abusers.[31] Sociologists have also begun to examine the treatment of animals in scientific experimentation.[32] We hope that this output will increase dramatically and that these issues will be examined more explicitly within the framework of social stratification.

In fond memory of Marquis (courtesy of Allen Landes).

Afterword

THE PHYSICAL SHELTER where we conducted our study no longer exists. Whiskers, however, continues to thrive and to rescue stray and abandoned cats. The shelter was moved to a two-story house, which allows for greater separation of the infirmary and the creation of several new special-purpose rooms, including a room for the elderly cats. Soon after the change in shelter location, Marquis was moved to this room when he could no longer fulfill his duties as shelter security officer because of old age. Some months later, he allowed himself to be adopted by one of the long-time volunteers, and he lived happily with her until his death in the fall of 2001. Although no one knew for certain, we believe that he was about 16 years old when he died.

Moving 80-odd cats to a new location must have been quite a scene, which we unfortunately missed because we were out of town. We have not yet investigated whether the new setting has changed the social relationships between the volunteers and the cats or among the cats. Kate told us that only Sid and Indigo took the move badly and that most of the cats got right into the spirit of things. Thus, we are hopeful that the special nature of Whiskers remains unchanged.

For more information about Whiskers, please go to their website: <http://www.albany.net/~qcats/index1.html>.

Notes

Preface

1. Hugh LaFollette and Niall Shanks, *Brute Science: Dilemmas of Animal Experimentation* (New York: Routledge, 1996), 267–68.

Chapter 1

1. Katharine M. Rogers, *The Cat and the Human Imagination: Feline Images from Bast to Garfield* (Ann Arbor: University of Michigan Press, 1998), 3.
2. Quoted in Rogers 1998, 89.
3. Rogers 1998, 87.
4. See Rogers 1998; see also James M. Jaspers and Dorothy Nelkin, *The Animal Rights Crusade: The Growth of a Moral Protest* (New York: Free Press, 1992).
5. Mark Twain, *Pudd'nhead Wilson* (London: Zodiac Press, 1967 [1894]), 33.
6. Leila Usher, "I Am the Cat," pp. 590–91 in *The Best Loved Poems of the American People*, ed. Hazel Felleman (Garden City, N.Y.: Doubleday, 1936).
7. Rogers 1998, 131.
8. Rogers 1998, 117.
9. Rogers 1998, 134.
10. Rogers 1998, 136.
11. Eric Swanson, *Hero Cats: True Stories of Daring Feline Deeds* (Kansas City, Mo.: Andrews McMeel, 1998), 10.
12. Swanson 1998, 22.

13. Swanson 1998, 33.

14. Swanson 1998, 94–98.

15. Annie, one of our recently deceased cats, was feral. She was very timid with people she did not know and with the other cats. Shelly, whom we acquired in 1994, took an intense dislike to Annie, and none of our efforts to change this situation were successful. Shelly spent time with Calvin, Cassidy, and Sydney. She seemed to have taught them to hate Annie, because none of them initially had a strong reaction to her. It was only after several months of observing Shelly's attitude and behavior toward Annie that these cats turned hostile to her. Jennifer, who did not associate with Shelly, never showed hostility toward her. We mention this because Annie would have loved to be with all of us but was prevented from doing so by the gang of four. She had to be very careful of them, because they rarely allowed her to sit peacefully on the main floor. They tolerated her to a greater extent on the upper floor, which they used only occasionally. We reinforced this situation because we wanted Annie to have someplace she could be in peace. Since she always slept with us, the upper floor was best.

16. Janet M. Alger and Steven F. Alger, "Beyond Mead: Symbolic Interaction between Humans and Felines," *Society and Animals* 5 (1997): 65–81.

17. Mead is laboring under what is called the Cartesian model of animal behavior (after the seventeenth-century philosopher Rene Descartes), which has been largely discredited by more recent reinterpretations of animal behavior research. This undermining of the Cartesian paradigm calls upon sociology to look again at the line that separates humans from other animals and to reexamine the central ideas of symbolic interactionism in the light of this new evidence. There are many elements in Meadian thought that are compatible with the new animal research if one does not focus on language as the central mechanism through which a self emerges or see consciousness of self as a primarily cognitive process.

18. Charles Horton Cooley, *Human Nature and the Social Order* (New York: Schocken Books, 1964), 196.

19. See, for instance, M. Donaldson, *Children's Minds* (New York: W. W. Norton, 1979).

20. Randall Collins, "Toward a Neo-Meadian Sociology of Mind," *Symbolic Interaction* 12 (1989): 1–32. Collins argues that Mead dwelled almost exclusively on these types of issues in his development of symbolic interactionism.

21. Collins 1989, 17–18.

22. Donald R. Griffin, *Animal Minds* (Chicago: University of Chicago Press, 1992).

23. See, for instance, Arnold Arluke and Clinton R. Sanders, *Regarding Animals* (Philadelphia: Temple University Press, 1996); Clinton R. Sanders, "Understanding Dogs: Caretakers' Attributes of Mindedness in Canine-Human Relationships," *Journal of Contemporary Ethnography* 22 (1993): 205–26; Clinton R. Sanders, *Understanding Dogs: Living and Working with Canine Companions* (Philadelphia: Temple University Press, 1999); Clinton R. Sanders and Arnold Arluke, "If Lions Could Speak: Investigating the Animal-Human Relationship and the Perspectives of Nonhuman Others," *Sociological Quarterly* 34 (1993): 377–90.

24. Sanders 1993, 210–11.

25. Alger and Alger 1997, 74.

26. From a letter sent to us by one of the respondents right after the interview telling us about several things she did not think of during the interview. She adopted Piccolina from Whiskers several years ago.

27. All quotations are from interviews.

28. Alger and Alger 1997, 75.

29. Quotations from interviews.

30. Alger and Alger 1997, 75.

31. Quotation from interviews.

32. First quotation from interviews. Others from Alger and Alger 1997, 76.

33. Quotations from interviews.

34. Sanders 1993.

35. Sanders 1993, 217.

36. Quotations from interviews.

37. Alger and Alger 1997, 76.

38. Quotation from interviews.

39. Alger and Alger 1997, 77.

40. Quotation from interviews.

41. Alger and Alger 1997, 78.

42. Quotation from interviews.

43. Collins 1989.

44. Arluke and Sanders 1996, 79.

45. L. A. Dugatkin and R. C. Sargent, "Male-Male Association Patterns and Female Proximity in the Guppy, Poecilia Reticulata," *Behavioral Ecology and Sociobiology* 35 (1994): 141–45.

Chapter 2

1. We present the historical information about Whiskers from interviews with the copresidents and from our own recollections and records as active participants in the reorganization of the shelter in the early 1990s.

2. The term "collector" refers to someone who takes in more cats (or other animals) than he or she can care for.

3. This undated memo was the first sent to the volunteers after the reorganization and was written sometime in July 1991. It is from our Whiskers file.

4. The term "feral cat" refers to a cat born and reared apart from humans. See Chapter 5 for a more extended discussion.

5. Arnold Arluke and Clinton R. Sanders, *Regarding Animals* (Philadelphia: Temple University Press, 1996), 19.

6. Burke Forrest, "Apprentice-Participation: Methodology and the Study of Subjective Reality," *Urban Life* 14 (1986): 431–53.

7. William F. Whyte, *Street Corner Society* (Chicago: University of Chicago Press, 1967), 305.

8. Whyte 1967, 298–99.

9. Forrest 1986.

10. Patricia A. Adler and Peter Adler, *Membership Roles in Field Research* (Beverly Hills, Calif.: Sage, 1987).

11. Adler and Adler 1987, 67.

12. Arluke and Sanders 1996.

13. Marian Stamp Dawkins, "Minding and Mattering," pp. 151–60 in *Mindwaves: Thoughts on Intelligence, Identity and Consciousness,* ed. Colin Blakemore and Susan Greenfield (Oxford: Basil Blackwell, 1987).

14. Marian Stamp Dawkins, *Through Our Eyes Only? The Search for Animal Consciousness* (Oxford: Oxford University Press, 1998), 155–56.

15. Roger Tabor, *The Wild Life of the Domestic Cat* (London: Arrow Books, 1983).

16. See, for instance, Martyn Hammersley, "What's Wrong with Ethnography? The Myth of Theoretical Description," *Sociology* 24 (1990): 597–615; Thomas J. Scheff, "Toward Resolving the Controversy over 'Thick Description,'" *Current Anthropology* 27 (1986): 408–9.

17. Heather Busch and Burton Silver, *Why Cats Paint: A Theory of Feline Aesthetics* (Berkeley, Calif.: Ten Speed Press, 1994).

18. In this respect our fieldwork role comes closest to what the Adlers (1987) call the "opportunistic complete member researcher" and what others call "autoethnography." See David M. Hayano, "Auto-Ethnography: Paradigms, Problems, and Prospects," *Human Organization* 38 (1979): 99–104.

19. Hammersley 1990.

20. Clifford Geertz, *The Interpretation of Cultures* (New York: Basic Books, 1973), 27.

21. Quoted in Arluke and Sanders 1996, 43.

22. Elizabeth Marshall Thomas, *The Tribe of Tiger: Cats and Their Culture* (New York: Simon & Schuster, 1994), 109.

23. Thomas 1994, 110.

24. Hammersley 1990.

Chapter 3

1. This not a whole number because a few cleaner/feeders commit to a shift every other week rather than every week.

2. David Horton Smith, "Altruism, Volunteers and Volunteering," *Journal of Voluntary Action Research* 10 (1981): 33.

3. Gary Fine, "Negotiated Orders and Organizational Cultures," *Annual Review of Sociology* 10 (1984): 239–62.

4. Fine 1984, 243.

5. Janet M. Alger and Steven F. Alger, "Cat Culture, Human Culture: An Ethnographic Study of a Cat Shelter," *Society and Animals* 7 (1999): 199–218.

6. Quoted in Herbert Blumer, *Symbolic Interactionism: Perspective and Method* (Berkeley: University of California Press, 1969), 61.

7. Charles Horton Cooley, *Human Nature and the Social Order* (New York: Schocken Books, 1964).

8. Robert Bogdan and S. J. Taylor, "Relationships with Severely Disabled People: The Social Construction of Humanness," *Social Problems* 36 (1989): 136.

9. Susan D. Suarez and G. G. Gallup, Jr., "Self-Recognition in Chimpanzees and Orangutans, but not Gorillas," *Journal of Human Evolution* 10 (1981): 175–88. The case of monkeys is actually controversial. In 1995, Hauser and colleagues published a paper in which they argued that they were successful in obtaining mirror recognition with cotton-top tamarins. Gallup and his colleagues do not believe the test was successful. For both sides of the argument see M. D. Hauser and J. Kralik, "Life beyond the Mirror: A Reply to

Anderson and Gallup," *Animal Behavior* 54 (1997): 1568–71; J. R. Anderson and G. G. Gallup, Jr., "Self-Recognition in *Saguinus*? A Critical Essay," *Animal Behavior* 54 (1997): 1563–67.

10. Even though such studies may be replicated, the total number of animals studied is still small.

11. Suarez and Gallup 1981, 186.

12. Gordon G. Gallup Jr., Michael K. McClure, Suzanne D. Hill, and Rosalie A. Bundy, "Capacity for Self-Recognition in Differentially Reared Chimpanzees," *The Psychological Record* 21 (1971): 73–74.

13. Quoted in Natalie Angier, "Evolutionary Necessity or Glorious Accident? Biologists Ponder the Self," *New York Times*, 22 April 1997, C9.

14. Quoted in Angier 1997, C1.

15. Quoted in Elizabeth Pennisi, "Are Our Primate Cousins 'Conscious'?" *Science* 284 (1999): 2076.

16. Antonio Damasio, *The Feeling of What Happens: Body and Emotion in the Making of Consciousness* (New York: Harcourt, Brace, 1999), 4.

17. Laura Tangley, "Animal Emotions: Sheer Joy. Romantic Love. The Pain of Mourning. Scientists Say Pets and Wild Creatures have Feelings, Too," *U.S. News & World Report*, 30 October 2000, 48–52.

18. Quoted in Angier 1997, C9. See also Stephen W. Porges, "The Polyvagal Theory: Phylogenetic Substrates of a Social Nervous System," *International Journal of Psychophysiology*, 42 (2): 29–52.

19. Donald R. Griffin, *Animal Minds* (Chicago: University of Chicago Press, 1992); Marian Stamp Dawkins, *Through Our Eyes Only? The Search for Animal Consciousness* (Oxford: Oxford University Press, 1998).

20. Dawkins 1998, 20.

21. Dawkins 1998, 36.

22. Dawkins 1998, 45.

23. Dawkins 1998, 53.

24. Dawkins 1998, 58–61.

25. Bogdan and Taylor 1989, 141–43.

26. Bogdan and Taylor 1989, 142–43.

27. Marc Bekoff, *The Smile of a Dolphin: Remarkable Accounts of Animal Emotions* (New York: Discovery Books, 2000), 21.

28. E. D. McCarthy, "Emotions are Social Things: An Essay in the Sociology of Emotions," p. 57 in *The Sociology of Emotions:*

Original Essays and Research Papers, ed. E. D. McCarthy and D. D. Franks (Greenwich, Conn.: JAI Press, 1989).

29. P. A. Thoits, "The Sociology of Emotions," *Annual Review of Sociology*, 15 (1989): 324.

30. Kate demonstrated such empathy between volunteers and cats. Her special bond with cats is mentioned at several points in the book. Suffice it to say that Kate can do things with the cats that no one else can. She has a magic touch with the cats, and they regularly sought her out for attention.

Chapter 4

1. Victoria L. Voith and Peter L. Borchelt, "Social Behavior of Domestic Cats," pp. 248–56 in *Readings in Companion Animal Behavior*, ed. Victoria L. Voith and Peter L. Borchelt (Trenton, N.J.: Veterinary Learning Systems Co., Inc., 1996).

2. See, for example, Theodore Xenophon Barber, *The Human Nature of Birds: A Scientific Discovery with Startling Implications* (New York: St. Martin's, 1993); Donald R. Griffin, *The Question of Animal Awareness: Evolutionary Continuity of Mental Experience* (New York: Rockefeller University, 1976); Donald R. Griffin, *Animal Thinking* (Cambridge, Mass.: Harvard University Press, 1984); Donald R. Griffin, *Animal Minds* (Chicago: University of Chicago Press, 1992). See also Voith and Borchelt 1996.

3. Rodney Stark, *Sociology* (Belmont, Calif.: Wadsworth, 1992), 141–42.

4. P. Marler and M. Tamura, "Song 'Dialects' in Three Populations of White-Crowned Sparrows," *Science* 146 (1964): 1483–86.

5. A. Whiten et al., "Cultures in Chimpanzees," *Nature* 399 (1999): 682–85.

6. Natalie Angier, "Chimps Exhibit er, Humanness, Study Finds," *New York Times* [on-line], 18 June 1999. Available: <http://www.nytimes.com/library/national/science/061799sci-animals-chimpanzee.html>.

7. See, for example, J. T. Bonner, *The Evolution of Culture in Animals* (Princeton, N.J.: Princeton University Press, 1980); Jane Goodall, *The Chimpanzees of Gombe: Patterns of Behavior* (Cambridge, Mass.: Belknap, 1986).

8. Marian Stamp Dawkins, *Through Our Eyes Only? The Search for Animal Consciousness* (Oxford: Oxford University Press, 1998), 45–53.

9. See, for example, Bennett G. Galef, "The Question of Animal Culture," *Human Nature* 3 (2): 157–78.

10. Frans B. M. de Waal, "Cultural Primatology Comes of Age," *Nature* 399 (1999): 636.

11. See, for example, Roger Tabor, *The Wild Life of the Domestic Cat* (London: Arrow Books, 1983); Elizabeth Marshall Thomas, *The Tribe of Tiger: Cats and Their Culture* (New York: Simon & Schuster, 1994). See also Voith and Borchelt 1996.

12. Thomas 1994, 109–10.

13. Thomas 1994, 110–12.

14. Janet M. Alger and Steven F. Alger, "Cat Culture, Human Culture: An Ethnographic Study of a Cat Shelter," *Society and Animals* 7 (1999): 199–218.

15. Paul Leyhausen, *Cat Behavior* (London: Garland STPM Press, 1979).

16. Tabor 1983, 57.

17. Tabor 1983, 75–76.

18. As we discuss in Chapter 5, a solid core of the old-timers were feral cats who had a distinctive impact on the conviviality of the cage tops.

19. Voith and Borchelt 1996.

20. For instance, between October 7, 1998, and January 30, 2000, we have 24 observations of them sleeping together. During that same period, Merlin slept with other cats on 8 occasions and Corey slept with other cats on 18 occasions. Of course, we are not at the shelter all of the time and, thus, cannot know the significance of these figures.

21. Tabor 1983, 75. See also G. Kerby and D. W. MacDonald, "Cat Society and the Consequences of Colony Size," p. 80 in *The Domestic Cat: The Biology of Its Behaviour*, ed. Dennis C. Turner and Patrick Bateson (Cambridge: Cambridge University Press, 1988).

22. A. H. Maslow, *Motivation and Personality* (New York: Harper and Row, 1970).

23. Janet M. Alger and Steven F. Alger, "Beyond Mead: Symbolic Interaction between Humans and Felines," *Society and Animals* 5 (1997): 65–81.

24. Tabor 1983.

25. Leyhausen 1979, 224.

26. Leyhausen 1979, 232–43.

27. Tabor 1983, 74–75.

28. Heather Busch and Burton Silver, *Why Cats Paint: A Theory of Feline Aesthetics* (Berkeley: Ten Speed Press, 1994).
29. Voith and Borchelt 1996, 249.
30. Voith and Borchelt 1996, 249–50.
31. Eric Swanson, *Hero Cats: True Stories of Daring Feline Deeds* (Kansas City, Mo.: Andrews McMeel, 1998), 39–47.
32. Leyhausen 1979, 232–43.
33. A "time-out" cage is not always available at the shelter, but when an especially aggressive animal like Emilio is present, a cage in the living room is kept empty to house the aggressor.
34. Voith and Borchelt 1996.
35. Tabor 1983, 57.
36. Leyhausen 1979, 217.
37. Leyhausen 1979, 218–19.
38. Leyhausen 1979, 218–20; Voith and Borchelt 1996.
39. Leyhausen 1979, 217–26.
40. Leyhausen 1979: 217–26.
41. Tabor 1983, 57–82.
42. Theodore Xenophon Barber, *The Human Nature of Birds: A Scientific Discovery with Startling Implications* (New York: St. Martin's, 1993), 31.
43. Dawkins 1998.
44. Dawkins 1998, 46–53.

Chapter 5

1. See, for example, Merritt Clifton, "Seeking the Truth about Feral Cats and the People Who Help Them," *Animal People*, November 1992, 1ff.; Jack Couffer, *The Cats of Lamu* (New York: Lyons Press, 1998); Paul Leyhausen, *Cat Behavior* (London: Garland STPM Press, 1979); Roger Tabor, *The Wild Life of the Domestic Cat* (London: Arrow Books, 1983).
2. Clifton 1992, 1.
3. Clifton 1992, 1.
4. See Couffer 1998; Leyhausen 1979; Tabor 1983.

Chapter 6

1. We have already referred to the cat and her kittens who we found on campus as the origin of our relationship with Whiskers. Here we note that we adopted the mother (Jennifer) in January

about a month after bringing her to the shelter, but the kitten (Jesse) was too wild for anyone to adopt at that time. We left her at the shelter in the hope that we and Kate could make her more handleable and suitable for adoption. By August it became clear that she would never be an adoptable cat and that we were the only people who could handle her. We felt responsible for her and took her home, where we sequestered her in a room closed with a screen door so she and our other cats could see but not come in contact with one another. When her mother came to the screen for the first time to see who was there, Jesse became ecstatic. She cried out, rolled over, danced, and rubbed against the screen. We let her mother in. Jennifer seemed initially shocked to see Jesse but within a few minutes was washing her and participating in the tremendous greeting her daughter was giving her. From that time on they were inseparable and Jesse enjoyed the happiest time in her life since being captured. Jesse certainly did not respond to our other cats in this way when they approached the screen door. There is little doubt that the seven months of separation neither caused her to forget her mother nor lose her attachment for her. It is important to note that Jesse was the kitten that Jennifer kept with her as the others began to develop their own lives. They were together for four months in the wild and another month at Whiskers before they were separated. Thus, they had the time to develop a close bond. When we return from vacations, everyone turns out to give us a big greeting, and they immediately want to do the things they can only do when we are home–such as go out in their run. In addition, they are very dependent for two or three days, following us everywhere, sitting in our laps, and so on.

2. See Neil J. Smelser, "The Rational and the Ambivalent in the Social Sciences," *American Sociological Review* 63 (1998): 1–15. He provides an account of the significance of ambivalence in social life.

3. Bibbi and Lisa were feral cats who had adapted to shelter life. They were both rather inflexible and fragile. Betty and I were concerned that they would not fare well outside of the shelter and should only go to the home of a volunteer who could carefully monitor their reactions and bring them back if necessary. Megan, who handled the adoption, disagreed. The adopter, she said, had a very good track record with her last cat, who lived for 20 years, and that no one else wanted Bibbi and Lisa.

4. Whiskers experienced regular member loss by design. That is, the goal of the organization is adoption. We raise the issue of the

impact of such member loss on the human–cat community collectively. Although it seems likely that there are parallels in other organizations, we did not find anything comparable in the literature. We also were unable to obtain literature examples from experts in the field of organizations. Thus, we cannot place our findings in a broader sociological context.

Chapter 7

1. Marian Stamp Dawkins, *Through Our Eyes Only? The Search for Animal Consciousness* (Oxford: Oxford University Press, 1998).
2. Dawkins 1998, 68.
3. Emile Durkheim, *Suicide* (Glencoe, Ill.: Free Press, 1951), 310.
4. Ruth A. Wallace and Alison Wolf, *Contemporary Sociological Theory: Expanding the Classical Tradition*, 5th ed. (Upper Saddle River, N.J.: Prentice-Hall, 1999), 383. The quotation within the quotation is from a letter to the *New York Review of Books*, 13 November 1975, from the Cambridge, Massachusetts–based Sociobiology Study Group.
5. Herbert Blumer, *Symbolic Interactionism: Perspective and Method* (Berkeley: University of California Press, 1969).
6. Paul Leyhausen, *Cat Behavior* (London: Garland STPM Press, 1979).
7. G. Kerby and D. W. MacDonald, "Cat Society and the Consequences of Colony Size," pp. 70–71 in *The Domestic Cat: The Biology of Its Behaviour*, ed. Dennis C. Turner and Patrick Bateson (Cambridge: Cambridge University Press, 1988).
8. Roger Tabor, *The Wild Life of the Domestic Cat* (London: Arrow Books, 1983), 75–76.
9. A. H. Maslow, *Motivation and Personality* (New York: Harper and Row, 1970).
10. Dawkins 1998.
11. Marc Bekoff, ed. *The Smile of a Dolphin: Remarkable Accounts of Animal Emotions* (New York: Discovery Books, 2000).
12. Frank Bruni, "Why Did the Cat Cross the Road? Because New York State Law Is on Her Side Now," *New York Times*, 10 September 1997, B1.
13. Legislature of the State of New York, "An Act to Amend the Agriculture and Markets Law, in Relation to Aggravated Cruelty to Animals," *Laws of New York, 1999*, Chapter 118.

14. Jonathan Balcombe, "Dissection and the Law," *The AV Magazine* 1996 (Summer): 18–21.

15. Evelyn Nieves, "Saving Cats and Dogs, in a 'No-Kill' Nation," *New York Times* [on-line], 18 January 1999. Available: <http://www.nytimes.com/yr/mo/day/news/national/sanfran-spca.html>. See also The Humane Society of the United States, "HSUS Pet Overpopulation Estimates," 2002 [on-line] Available: <http://www.hsus.org/ace/11830>.

16. <http://www.vivisectioninfo.org/cat.html> and <http://www.vivisectioninfo.org/dog.html>.

17. Dawkins 1998, Chapter 5.

18. Janet M. Alger and Steven F. Alger, "The Reconstruction of Animals and the Redefinition of Animal Oppression," paper presented at the American Sociological Association Annual Meeting, Washington, D.C., August 12–16, 2000.

19. New York State passed such legislation in 1994. Assemblyman Pete Grannis introduced the bill. Our information comes from his office.

20. For comparable findings for dogs, see Clinton R. Sanders, *Understanding Dogs: Living and Working with Canine Companions* (Philadelphia: Temple University Press, 1999). See especially Chapter 6.

Chapter 8

1. Rick Scarce, "Socially Constructing Pacific Salmon," *Society and Animals* 5 (1997): 117–35.

2. Jennifer R. Wolch, Andrea Gullo, and Unna Lassiter, "Changing Attitudes toward California's Cougars," *Society and Animals* 5 (1997): 95–116.

3. Michael S. Quinn, "Corpulent Cattle and Milk Machines," *Society and Animals* 1 (1993): 145–57.

4. Lyle Mauro, "Framing Cruelty: The Construction of Duck Shooting as a Social Problem," *Society and Animals* 5 (1997): 137–54.

5. Lawrence G. Carbone, "Death by Decapitation: A Case Study of the Scientific Definition of Animal Welfare," *Society and Animals* 5 (1997): 239–56.

6. Shelly L. Galvin and Harold A. Herzog, Jr., "Attitudes and Dispositional Optimism of Animal Rights Demonstrators," *Society and Animals* 6 (1998): 1–11.

7. Lyle Munro, "Contesting Moral Capital in Campaigns against Animal Liberation," *Society and Animals* 7 (1999): 35–53.

8. Frank R. Ascione, Claudia V. Weber, and David S. Wood, "The Abuse of Animals and Domestic Violence: A National Survey of Shelters for Women who are Battered," *Society and Animals* 5 (1997): 205–18.

9. Carol D. Raupp, Mary Barlow, and Judith A. Oliver, "Perceptions of Family Violence: Are Companion Animals in the Picture?" *Society and Animals* 5 (1997): 219–37.

10. David A. Nibert, "Animal Rights and Human Social Issues," *Society and Animals* 2 (1994): 115–24.

11. Lynda Birke and Mike Michael, "The Heart of the Matter: Animal Bodies, Ethics, and Species Boundaries," *Society and Animals* 6 (1998): 245–61. Tania Woods, "Have a Heart: Xenotransplantation, Nonhuman Death and Human Distress," *Society and Animals* 6 (1998): 47–65.

12. Charles Horton Cooley, *Human Nature and the Social Order* (New York: Schocken Books, 1964).

13. Robert Bogdan and S. J. Taylor, "Relationships with Severely Disabled People: The Social Construction of Humanness," *Social Problems* 36 (1989): 135–48.

14. Susan D. Suarez and G. G. Gallup, Jr., "Self-Recognition in Chimpanzees and Orangutans, but not Gorillas," *Journal of Human Evolution* 10 (1981): 175–88.

15. Porges, Stephen W., "The Polyvagal Theory: Phylogenetic Substrates of a Social Nervous System," *International Journal of Psychophysiology* 42(2): 2.

16. Cooley 1964.

17. George Herbert Mead, *Mind, Self, and Society* (Chicago: University of Chicago Press, 1962).

18. Randall Collins, "Toward a Neo-Meadian Sociology of Mind," *Symbolic Interaction* 12 (1989): 1–32.

19. Cooley 1964, 196.

20. Antonio Damasio, *The Feeling of What Happens: Body and Emotion in the Making of Consciousness* (New York: Harcourt, Brace, 1999).

21. Damasio 1999, 30–31.

22. Damasio 1999, 107–8.

23. See, for example, Douglas S. Massey, "A Brief History of Human Society: The Origin and Role of Emotion in Social Life," *American Sociological Review* 67 (2002): 1–29.

24. E. D. McCarthy, "Emotions are Social Things: An Essay in the Sociology of Emotions," p. 57 in *The Sociology of Emotions. Original Essays and Research Papers,* ed. E. D. McCarthy and D. D. Franks (Greenwich, Conn.: JAI Press, 1989).

25. Damasio 1999, 53.

26. Damasio 1999, 53.

27. Marian Stamp Dawkins, *Through Our Eyes Only? The Search for Animal Consciousness* (Oxford: Oxford University Press, 1998): 70–71.

28. Peter Singer, *Animal Liberation,* new revised ed. (New York: Avon Books, 1990), 95.

29. Singer 1990, 123–25; Betsy Freese, "No Shortcut to Pig Welfare," *Successful Farming* 1990 (August): 22–24.

30. F. Barbara Orlans, "Data on Animal Experimentation in the United States: What They Do and Do Not Show," *Perspectives in Biology and Medicine* 37 (2): 217-31.

31. Janet M. Alger and Steven F. Alger, "The Reconstruction of Animals and the Redefinition of Animal Oppression," paper presented at the American Sociological Association Annual Meeting, Washington, D.C., August 12–16, 2000.

32. See, for example, Arnold Arluke, "Going into the Closet with Science: Information Control among Animal Experimenters," *Journal of Contemporary Ethnography* 20 (1991): 306–30. See also Mary Phillips, "Savages, Drunks, and Lab Animals: The Researcher's Perception of Pain," *Society and Animals* 1 (1993): 61–81.

References

Adler, Patricia A., and Peter Adler. 1987. *Membership Roles in Field Research.* Beverly Hills, Calif.: Sage.

Alger, Janet M., and Steven F. Alger. 1997. "Beyond Mead: Symbolic Interaction between Humans and Felines." *Society and Animals* 5: 65–81.

——. 1999. "Cat Culture, Human Culture: An Ethnographic Study of a Cat Shelter." *Society and Animals* 7: 199–218.

——. 2000. "The Reconstruction of Animals and the Redefinition of Animal Oppression." Presented at the American Sociological Association Annual Meeting, Washington, D.C., August 12–16.

Anderson, J. R., and G. G. Gallup, Jr. 1997. "Self-Recognition in *Saguinus*? A Critical Essay." *Animal Behavior* 54: 1563–67.

Angier, Natalie. 1997. "Evolutionary Necessity or Glorious Accident? Biologists Ponder the Self." *New York Times* 22 April: C1+.

——. 1999. "Chimps Exhibit er, Humanness, Study Finds." *New York Times.* [On-line]. 18 June. Available: <http://www.nytimes.com/library/national/science/061799sci-animals-chimpanzee.html>.

Arluke, Arnold. 1991. "Going into the Closet with Science: Information Control among Animal Experimenters." *Journal of Contemporary Ethnography* 20: 306–30.

Arluke, Arnold, and Clinton R. Sanders. 1996. *Regarding Animals.* Philadelphia: Temple University Press.

Ascione, Frank R., Claudia V. Weber, and David S. Wood. 1997. "The Abuse of Animals and Domestic Violence: A National Survey of Shelters for Women who are Battered." *Society and Animals* 5: 205–18.

Balcombe, Jonathan. 1996. "Dissection and the Law." *The AV Magazine* (Summer).

Barber, Theodore Xenophon. 1993. *The Human Nature of Birds: A Scientific Discovery with Startling Implications.* New York: St. Martin's.

Bekoff, Marc, ed. 2000. *The Smile of a Dolphin: Remarkable Accounts of Animal Emotions.* New York: Discovery Books.

Birke, Lynda, and Mike Michael. 1998. "The Heart of the Matter: Animal Bodies, Ethics, and Species Boundaries." *Society and Animals* 6: 245–61.

Blumer, Herbert. 1969. *Symbolic Interactionism: Perspective and Method.* Berkeley: University of California Press.

Bogdan, Robert, and S. J. Taylor. 1989. "Relationships with Severely Disabled People: The Social Construction of Humanness." *Social Problems* 36: 135–48.

Bonner, J. T. 1980. *The Evolution of Culture in Animals.* Princeton, N.J.: Princeton University Press.

Bruni, Frank. 1997. "Why Did the Cat Cross the Road? Because New York State Law Is on Her Side Now." *New York Times.* 10 September: B1.

Busch, Heather, and Burton Silver. 1994. *Why Cats Paint: A Theory of Feline Aesthetics.* Berkeley: Ten Speed Press.

Carbone, Lawrence G. 1997. "Death by Decapitation: A Case Study of the Scientific Definition of Animal Welfare." *Society and Animals* 5: 239–56.

Clifton, Merritt. 1992. "Seeking the Truth about Feral Cats and the People Who Help Them." *Animal People* November: 1ff.

Collins, Randall. 1989. "Toward a Neo-Meadian Sociology of Mind." *Symbolic Interaction* 12: 1–32.

Cooley, Charles Horton. 1964. *Human Nature and the Social Order.* New York: Schocken Books.

Couffer, Jack. 1998. *The Cats of Lamu.* New York: Lyons Press.

Damasio, Antonio. 1999. *The Feeling of What Happens: Body and Emotion in the Making of Consciousness.* New York: Harcourt, Brace.

Dawkins, Marian Stamp. 1987. "Minding and Mattering." Pp. 151–60 in *Mindwaves: Thoughts on Intelligence, Identity and Consciousness,* ed. Colin Blakemore and Susan Greenfield. Oxford: Basil Blackwell.

—. 1998. *Through Our Eyes Only? The Search for Animal Consciousness.* Oxford: Oxford University Press.

de Waal, Frans B. M. 1999. "Cultural Primatology Comes of Age." *Nature* 399: 635–36.

Donaldson, M. 1979. *Children's Minds.* New York: W. W. Norton.

Dugatkin, L. A., and R. C. Sargent. 1994. "Male-Male Association Patterns and Female Proximity in the Guppy, Poecilia Reticulata." *Behavioral Ecology and Sociobiology* 35: 141–45.

Durkheim, Emile. 1951. *Suicide.* Glencoe, Ill.: Free Press.

Fine, Gary. 1984. "Negotiated Orders and Organizational Cultures." *Annual Review of Sociology* 10: 239–62.

Forrest, Burke. 1986. "Apprentice-Participation: Methodology and the Study of Subjective Reality." *Urban Life* 14: 431–53.

Freese, Betsy. 1990. "No Shortcut to Pig Welfare." *Successful Farming* (August), 22–24.

Galef, Bennett G. 1992. "The Question of Animal Culture." *Human Nature* 3(2): 157–78.

Gallup, Gordon G., Jr., Michael K. McClure, Suzanne D. Hill, and Rosalie A. Bundy. 1971. "Capacity for Self-Recognition in Differentially Reared Chimpanzees." *The Psychological Record* 21: 69–74.

Galvin, Shelly L., and Harold A. Herzog, Jr. 1998. "Attitudes and Dispositional Optimism of Animal Rights Demonstrators." *Society and Animals* 6: 1–11.

Geertz, Clifford. 1973. *The Interpretation of Cultures.* New York: Basic Books.

Goodall, Jane. 1986. *The Chimpanzees of Gombe: Patterns of Behavior.* Cambridge, Mass.: Belknap.

Griffin, Donald R. 1976. *The Question of Animal Awareness: Evolutionary Continuity of Mental Experience.* New York: Rockefeller University.

—. 1984. *Animal Thinking.* Cambridge, Mass.: Harvard University Press.

—. 1992. *Animal Minds.* Chicago: University of Chicago Press.

Hammersley, Martyn. 1990. "What's Wrong with Ethnography? The Myth of Theoretical Description." *Sociology* 24: 597–615.

Hauser, M. D., and J. Kralik. 1997. "Life beyond the Mirror: A Reply to Anderson and Gallup." *Animal Behavior* 54: 1568–71.

Hauser, M. D., J. Kralik, C. Botto, M. Garrett, and J. Oser. 1995. "Self-Recognition in Primates: Phylogeny and the Salience of Species-Typical Traits. *Proceedings of the National Academy of Sciences* 92: 10811–14.

Hayano, David M. 1979. "Auto-Ethnography: Paradigms, Problems, and Prospects." *Human Organization* 38: 99–104.

The Humane Society of the United States. 2002. "HSUS Pet Overpopulation Estimates." [On-line] Available: <http://www.hsus.org/ace/11830>.

Jaspers, James M., and Dorothy Nelkin. 1992. *The Animal Rights Crusade: The Growth of a Moral Protest.* New York: Free Press.

Kerby, G., and D. W. MacDonald. 1988. "Cat Society and the Consequences of Colony Size." Pp. 67–81 in *The Domestic Cat: The Biology of Its Behaviour,* ed. Dennis C. Turner and Patrick Bateson. Cambridge: Cambridge University Press.

LaFollette, Hugh, and Niall Shanks. 1996. *Brute Science: Dilemmas of Animal Experimentation.* New York: Routledge.

Leyhausen, Paul. 1979. *Cat Behavior.* London: Garland STPM Press.

Marler, P., and M. Tamura. 1964. "Song 'Dialects' in Three Populations of White-Crowned Sparrows." *Science* 146: 1483–86.

Maslow, A. H. 1970. *Motivation and Personality.* New York: Harper and Row.

Massey, Douglas S. 2002. "A Brief History of Human Society: The Origin and Role of Emotion in Social Life." *American Sociological Review* 67: 1–29.

Mauro, Lyle. 1997. "Framing Cruelty: The Construction of Duck Shooting as a Social Problem." *Society and Animals* 5: 137–54.

McCarthy, E. D. 1989. "Emotions are Social Things: An Essay in the Sociology of Emotions." Pp. 51–72 in *The Sociology of Emotions: Original Essays and Research Papers,* ed. E. D. McCarthy and D. D. Franks. Greenwich, Conn.: JAI Press.

Mead, George Herbert. 1962. *Mind, Self, and Society.* Chicago: University of Chicago Press.

Munro, Lyle. 1999. "Contesting Moral Capital in Campaigns against Animal Liberation." *Society and Animals* 7: 35–53.

Nibert, David A. 1994. "Animal Rights and Human Social Issues." *Society and Animals* 2: 115–24.

Nieves, Evelyn. 1999. "Saving Cats and Dogs, in a 'No-Kill' Nation." *New York Times.* [On-line]. 18 January. Available: <http://www.nytimes.com/yr/mo/day/news/national/sanfran-spca.html>.

Orlans, F. Barbara. 1994. "Data on Animal Experimentation in the United States: What They Do and Do Not Show." *Perspectives in Biology and Medicine* 37(2): 217–31.

Pennisi, Elizabeth. 1999. "Are Our Primate Cousins 'Conscious'?" *Science* 284: 2073–76.

Phillips, Mary. 1993. "Savages, Drunks, and Lab Animals: The Researcher's Perception of Pain." *Society and Animals* 1: 61–81.

Porges, Stephen W. 2001. "The Polyvagal Theory: Phylogenetic Substrates of a Social Nervous System." *International Journal of Psychophysiology* 42(2): 29–52.

Quinn, Michael S. 1993. "Corpulent Cattle and Milk Machines." *Society and Animals* 1: 145–57.

Raupp, Carol D., Mary Barlow, and Judith A. Oliver. 1997. "Perceptions of Family Violence: Are Companion Animals in the Picture?" *Society and Animals* 5: 219–37.

Rogers, Katharine M. 1998. *The Cat and the Human Imagination: Feline Images from Bast to Garfield.* Ann Arbor: University of Michigan Press.

Sanders, Clinton R. 1993. "Understanding Dogs: Caretakers' Attributes of Mindedness in Canine-Human Relationships." *Journal of Contemporary Ethnography* 22: 205–26.

——. 1999. *Understanding Dogs: Living and Working with Canine Companions.* Philadelphia: Temple University Press.

Sanders, Clinton R., and Arnold Arluke. 1993. "If Lions Could Speak: Investigating the Animal-Human Relationship and the Perspectives of Nonhuman Others." *Sociological Quarterly* 34: 377–90.

Scarce, Rick. 1997. "Socially Constructing Pacific Salmon." *Society and Animals* 5: 117–35.

Scheff, Thomas J. 1986. "Toward Resolving the Controversy over 'Thick Description.'" *Current Anthropology* 27: 408–9.

Singer, Peter. 1990. *Animal Liberation.* New revised ed. New York: Avon Books.

Smelser, Neil J. 1998. "The Rational and the Ambivalent in the Social Sciences." *American Sociological Review* 63: 1–15.

Smith, David Horton. 1981. "Altruism, Volunteers and Volunteering." *Journal of Voluntary Action Research* 10: 21–36.

Stark, Rodney. 1992. *Sociology.* Belmont, Calif.: Wadsworth.

Suarez, Susan D., and G. G. Gallup, Jr. 1981. "Self-Recognition in Chimpanzees and Orangutans, but not Gorillas." *Journal of Human Evolution* 10: 175–88.

Swanson, Eric. 1998. *Hero Cats: True Stories of Daring Feline Deeds.* Kansas City, Mo.: Andrews McMeel.

Tabor, Roger. 1983. *The Wild Life of the Domestic Cat.* London: Arrow Books.

Tangley, Laura. 2000. "Animal Emotions: Sheer Joy. Romantic Love. The Pain of Mourning. Scientists Say Pets and Wild Creatures have Feelings, Too." *U.S. News & World Report.* 30 October: 48–52.

Thoits, P. A. 1989. "The Sociology of Emotions." *Annual Review of Sociology* 15: 317–42.

Thomas, Elizabeth Marshall. 1994. *The Tribe of Tiger: Cats and Their Culture.* New York: Simon & Schuster.

Twain, Mark. 1967. *Pudd'nhead Wilson.* London: Zodiac Press. Original edition, 1894.

Usher, Leila. 1936. "I Am the Cat." Pp. 590–91 in *The Best Loved Poems of the American People,* ed. Hazel Felleman. Garden City, N.Y.: Doubleday.

Voith, Victoria L., and Peter L. Borchelt. 1996. "Social Behavior of Domestic Cats." Pp. 248–56 in *Readings in Companion Animal Behavior,* ed. Victoria L. Voith and Peter L. Borchelt. Trenton, N.J.: Veterinary Learning Systems Co., Inc.

Wallace, Ruth A., and Alison Wolf. 1999. *Contemporary Sociological Theory: Expanding the Classical Tradition.* 5th ed. Upper Saddle River, N.J.: Prentice-Hall.

Whiten, A., et al. 1999. "Cultures in Chimpanzees." *Nature* 399: 682–85.

Whyte, William F. 1967. *Street Corner Society.* Chicago: University of Chicago Press.

Wolch, Jennifer R., Andrea Gullo, and Unna Lassiter. 1997. "Changing Attitudes toward California's Cougars." *Society and Animals* 5: 95–116.

Woods, Tania. 1998. "Have a Heart: Xenotransplantation, Nonhuman Death and Human Distress." *Society and Animals* 6: 47–65.

Index

Italic page numbers indicate photographs.